NIMBU SAAB

NIMBU SAAB

THE BAREFOOT NAGA KARGIL HERO

NEHA DWIVEDI
DIKSHA DWIVEDI
WITH NEINGUTOULIE KENGURÜSE

HarperCollins *Publishers* India

First published in India by HarperCollins *Publishers* 2024
4th Floor, Tower A, Building No. 10, Phase II, DLF Cyber City,
Gurugram, Haryana – 122002
www.harpercollins.co.in

2 4 6 8 10 9 7 5 3 1

Copyright © Diksha Dwivedi and Neha Dwivedi 2024

P-ISBN: 978-93-6213-368-7
E-ISBN: 978-93-6213-225-3

The views and opinions expressed in this book are the authors' own
and the facts are as reported by them, and the publishers
are not in any way liable for the same.

Diksha Dwivedi and Neha Dwivedi assert the moral right
to be identified as the authors of this work.

All rights reserved. No part of this publication may be reproduced,
stored in a retrieval system, or transmitted, in any form or by any means,
electronic, mechanical, photocopying, recording or otherwise,
without the prior permission of the publishers.

Typeset in 11.5/17 Adobe Garamond at
Manipal Technologies Limited, Manipal

Printed and bound at
Thomson Press (India) Ltd

This book is produced from independently certified FSC® paper
to ensure responsible forest management.

For the people of the very warm and welcoming state of Nagaland, the officers and men of 2 Raj Rif and, most of all, the brave and very loving Kengurüse family.

Also, with hope, for every young person looking for a hero.

'A job appears dull, routine or mechanical only when it is looked at with dull eyes or a dull mind.'

—Capt. Neikezhakuo Kengurüse (Maha Vir Chakra recipient)

Contents

Authors' note ix
Prologue xv

1. Sense of belonging 1
2. God's plan 17
3. Men don't cry 31
4. Three minus one 45
5. The grief 59
6. My brother Neibu and me 75
7. Naga GC creates history 97
8. The rise of Lt Neikezhakuo Kengurüse 125
9. 'Your own ease, comfort and safety come last' ... 143
10. Bare feet feel the earth's heartbeat 165

Epilogue 187
A note from Neingutoulie Kengurüse 203
Acknowledgements 209
Notes 217

Authors' Note

To say that writing this book has been an enlightening and fulfilling journey would be an understatement. We are so much richer for it.

We will never forget the first day we saw Atoulie and his youngest siblings (out of thirteen), Neichünuo and Viketoulie at the Kargil War memorial in Drass. At the time, we only had a hunch that they may be the family members of Captain Neikezhakuo Kengurüse, the Maha Vir Chakra recipient. Looking at them, it seemed to us as if the younger ones were trying to hold on tight to their elder brother, their Neibu (Kengurüse's nickname at home). We couldn't stop ourselves from staring at them often during the twenty-second annual commemoration ceremony of Kargil Vijay Diwas, the event we were attending. We're glad we could eventually gather the courage to walk up to Atoulie (short for Neingutoulie) and ask him if they were indeed related to Capt. Kengurüse. We wanted to express to him how we had read every single one of his brother's letters, previously published in Diksha's *Letters from Kargil* (2017), and how sorry we were that his story has not been told yet.

Just then, before we could say any of it, Capt. Kengurüse's regimental officer, Brig. Mohit Saxena, walked in and joined our conversation. Teary-eyed, he said to Atoulie, 'None of them from the attack survived; I sometimes feel like it would have been better had I not come back either.' Atoulie immediately replied, 'No!

Authors' Note

Don't say that! Some of us have to survive to tell the stories of those who cannot.'

In that moment, an unspoken deal was sealed between us and Atoulie, the wise, eldest brother of the Kengurüse family after Neibu. The journey from then to the present has transformed us from strangers to collaborators and now friends.

Neibu had lived up to the meaning of his name—giving 'great happiness' to his family while offering his duty to the nation. While twenty-five years have now passed since his supreme sacrifice (which is what earned him the Maha Vir Chakra), not many people know about Neibu and who he was beyond his uniform. We decided to fix this, along with Atoulie.

While we put together most pieces of the story with Atoulie's help and in his voice—as large parts of it had been narrated by Atoulie—we also relied on various interviews, which we conducted during an eight-day trip to Nagaland. There, we had the opportunity to read his letters and journals in person and spoke at length to many of Neibu's family members, colleagues and friends. Some of them had also been his students, which helped us understand how he was as a teacher as well.

Furthermore, we visited 2 Rajputana Rifles, the oldest rifle regiment in the Indian Army, in Barrackpore to meet with the men Neibu commanded. We also met and interviewed, offline and virtually, some of the officers in Delhi, whom he trained with during his time in the academy as well as a few who fought the Kargil War with him.

In addition to all the above, to make some of the larger events clearer to the reader, we relied on books (including our previous

Authors' Note

ones, *Letters from Kargil* and *Vijyant at Kargil*), journals and news articles, all of which have helped us paint a better picture of Nagaland's history and its current state, as well as the happenings of the Kargil War.

Both of us have tried our best to do justice to the story of Capt. Kengurüse and his state, Nagaland. However, should there be some mistakes, please know that it is unintentional, and we would be happy to correct it in further editions.

As daughters of a Kargil martyr, Maj. C.B. Dwivedi, posthumous Sena Medal recipient, we have made sincere efforts to bring the story of another hero from the same war to light, with as much authenticity as we could for it is as much our responsibility as it is our privilege to be in the position to do so.

Having said that, we are happy to present to you the extraordinary story of an ordinary boy, which we hope will inspire and comfort you in equal measure.

Prologue

*A*TOULIE HAD TO ALMOST pinch himself again to accept that this was really happening. This was supposed to be HIS time. He was supposed to start living his life on his terms. He had been a shoulder, and a good one at that, to his elder brother Neibu while he was in the Indian Military Academy—that's what he believed, and he had every reason to think so. Now was the chance for him to take over after getting commissioned … Only now, Neibu couldn't keep his side of the bargain. While Atoulie waited for his brother in uniform to take over the expenses of the big happy Naga family in Kohima, Neibu was already on his way to a land of no return.

Atoulie was gathering his bearings to stand in front of the defence minister of the country, George Fernandes, and ask for his family's rights. He had no other choice. In a few minutes, Atoulie and his father, Neiselie, would leave for the dinner with the then defence minister. They were still in awe of the award ceremony, for which they had flown down from Dimapur.

Earlier during that trip, they had felt so proud to be standing in front of the President of the country, while a room full of important-looking people, listening with rapt attention, heard about the actions of their brave son and brother, Capt. Neikezhakuo Kengurüse.

How proud and how utterly helpless.

This trip had brought with it many firsts for the father and son duo—their first time in the capital city, their first time staying in a five-star hotel, their first time navigating their way up an elevator and feeling truly lost (literally and metaphorically).

After the ceremony, Atoulie heard from someone that Indian Oil had allotted a hundred petrol pumps to the families of Kargil war martyrs. But there were 527 martyrs. Had they been allotted one? How could he know if his family had? Even though his brother had sacrificed his life for the country (and in a way, his whole family's too), was it enough to be able to qualify for the allotment?

Just as Atoulie was looking for answers to all these questions in his head, his attention was suddenly drawn to what he had come all the way for—precisely 1,480 miles—from Nagaland to the capital city, New Delhi. Almost as if some higher power were in play. He looked down at the small yet heavy, maroon velvet box that he had been holding for his father. The small, ordinary-looking box carried a rather large load: The Maha Vir Chakra, the second highest military decoration in the country, which had been earned by his brother, posthumously. Atoulie needed nothing else to make his mind up. The battle on the field was over, but the one of their lives was just beginning, and he knew that his brother had done more than enough to give his family, the Kengurüses, a fighting chance at a comfortable lifestyle—a lifestyle most of them could never imagine without Neibu's audacious ambition. Now, Atoulie had to take his brother's dream forward.

As the time for the dinner came closer, Atoulie could feel a familiar nervousness, like one feels before taking an exam. However, once they stood at the door of the house of the honourable defence minister, he felt it melting away. Minister George Fernandes stood there himself, humbly welcoming all of them with a song he was singing for his special guests, the next of kin of the bravehearts of his country. As they all took small steps inside, the sound of *Que sera sera; whatever will be, will be*, in the voice of their gracious host, enveloped them like a warm, comforting blanket.

The atmosphere thus assumed a relaxed tone while they all mingled with each other. Atoulie and his father felt further relaxed once they met with a sweet, smart and pleasant young lady, whose name tag read 'Capt. Indu'. As the evening progressed, she made sure that they were comfortable and were never left feeling alone. She accompanied them everywhere, politely seeking permission to sit with them or eat with them. Perhaps it was because of her that Atoulie felt comfortable enough to encourage, and accompany, his father in eating the food, which was specially laid out for them, with his hands—instead of using the cutlery placed before them, which they did not know how to use.

By the time the meal was over, Capt. Indu had established a good rapport with the Kengurüses. Thus, when the defence minister invited anyone who wanted to speak to him to come up, Atoulie promptly went over to Capt. Indu and asked her to take him to Minister Fernandes.

As they were walking up to him, Atoulie felt a quiet sense of confidence surge within him. Being the brother of a brave Maha Vir Chakra awardee brought out the courage in him. When they sat down, Atoulie turned towards the defence minister, looked calmly into his eyes and said, 'Sir, we are so thankful to you for bringing us together here. Thank you for the honour that you have given us here. But I have a problem that I need to share.'

'Please share,' Fernandes said encouragingly.

Atoulie continued, 'Sir, I have ten brothers and sisters after me. Neibu was the only breadwinner of the family, but he's no more. Now, I have to take care of my younger ones. My father and mother are more or less illiterate. So, if you could give me one petrol pump, with that, I will definitely be able to look after my family, and I will be grateful to you throughout my life.'

The defence minister smiled and said, 'Okay, I will do something.' With that, he asked for a paper and pen and wrote a note addressed to the then minister of petroleum and natural gas in India, Ram Naik.

On his trip back to Nagaland, Atoulie almost couldn't believe that he had not only stood in front of the highest officials of the country but had also managed to point blank ask for his family's rights. Was this the promise that he had to fulfil all along? Was this God's plan?

Atoulie's thoughts took him back to the time when the promise was made ...

After finishing his graduation, Neibu had been working as a teacher of mathematics and science in a school, while also offering tuition classes alongside to make sure he could earn extra money to be able to support all his siblings and their pursuits too. Even though he was making a reasonably good amount every month, considering that his family included ten siblings aside from Atoulie, money was never enough.

Prologue

However, the more Neibu worked, the more he realized that his dreams were bigger than just surviving. He wanted to thrive. He didn't just want them to be able to make ends meet. In fact, he wanted to uplift the status of his family in society and make sure that each of his siblings thrived. The goal was big and ambitious, and as the only earning one in his family, he knew that it was up to him to work towards achieving it.

Tough as it was, Neibu never stopped working towards his dream. Which is why, even though his schedule, which was largely made up of teaching different groups of students and classes throughout the day, didn't leave him much time to invest on himself, he still managed to keep doing it.

Neibu managed to prepare for a few qualifying exams and took them when he could. His aim was to become an officer. A man with authority and standing in society. Someone whose position would command respect, and that would extend to his family, too.

When the results were out, Neibu had cleared not just one exam; he had qualified in three! One of the exams he cleared would allow him to qualify as a bank provisional officer; another would give him a job in life insurance; and the last was none other than the Combined Defence Services (CDS) exam to join the Indian Army.

This presented a real dilemma for him. Neibu spent the next few days meeting with and seeking the advice of elders in his family and community. He wanted to understand what, according to everyone or most people, was his best option. While everyone was happy to know about his options, most shared one belief in common. Joining the Indian Army was out of the question. For years people of his state and community had been caught between the Indian Army and, what some would call, Naga freedom fighters. Growing up witnessing, listening to and sometimes becoming collateral damage in the ongoing conflict had left many in the state wary of the Indian Army. Hence, that option had the least takers among the people Neibu had turned to for advice.

But there were some people who did encourage him to pursue a military career. One uncle, for instance, whose house he had briefly lived in during college—Kruzecha Rio, his paternal aunt's husband. His mentors, a public relations officer, Col Robertson, and Col J.G.K. Kutty, who had also helped him prepare for the exam, were among the others who encouraged him. Some were thus elated to hear the news and enthusiastically encouraged him to go for it. Whatever the reactions he received may have been, the idea of donning the uniform and becoming a decorated officer, with stars on his shoulders, was exactly how Neibu had envisioned uplifting the status of his family. He couldn't think of any other position that could command so much respect for him and his family.

Despite his inclination, he was concerned about his family's finances as none of his father's children, aside from him, were earning at the time. Atoulie, who was the third child of their father, was in college and hadn't started earning yet. Neibu thus did not want any of them to have to compromise on their studies in order to start earning for the family. Besides, none of them were yet equipped to earn enough either—enough considering the number of people in the family. He did not want any one of them to have the life he had been leading, with the pressure of it all.

Opting for the army would require him to first go for training, where he wouldn't get paid much aside from a basic stipend, which would barely be enough for his own needs, let alone taking care of his entire family. For a year and a half, which was the duration of the training that he was supposed to undergo, what would his family do?

Atoulie had been observing Neibu in the week after his results and could see that beneath the obvious happiness of cracking the exams, his brother was deeply worried; yet he wasn't sharing anything with him. But who could understand Neibu better than Atoulie? He knew exactly what was wrong. So, one morning, he just got up and went to Neibu's room, sat him down and made him a promise that empowered his elder brother to take the risk he wanted to and follow his heart.

When he found out that Neibu was finally returning from the academy, after one and a half years of training, before he joined the unit he was commissioned in, Atoulie was perhaps the most excited. He had fulfilled the promise that he had made to his brother, and finally it was time to reap the reward.

Thus, when he met Neibu after his passing out parade, he asked for his help with something he had been waiting to buy; he had put it on hold and had decided to do it only after Neibu had finished his training. Like Neibu, he too thought that his testing times were over.

The thing that Atoulie wanted to buy was a colour television, as he loved watching movies and felt he had worked hard enough to deserve one. Besides, there were few sources of entertainment in the village where he worked, so he often got bored and believed that having a TV would really take care of his boredom. He had been waiting for his brother to return, since he couldn't afford to spend on such expensive recreational activities by himself. Now the trying times seemed to have passed, and so Atoulie had finally, confidently, asked Neibu for some money to buy a colour television set.

When Neibu refused, Atoulie was taken aback. Atoulie wasn't even asking him to cover the whole cost. He had just asked for the difference that he needed since he had also saved up a little. But Neibu wouldn't hear of it and simply turned him away. What made matters worse was that Atoulie found out that Neibu had unflinchingly given a slightly smaller amount to one of their sisters, who had asked for it to buy a dress. Why did she deserve a gift from Neibu if Atoulie didn't?

Atoulie was not just disappointed. He was hurt. Unfortunately, that hurt gave way to some resentment too.

As a result, Atoulie stopped interacting with his brother. When Neibu would write letters to him after this incident, they would go unanswered. If and when they had an opportunity for a conversation, they would run out of words. However, Neibu didn't stop.

Neibu continued to write letters to him as he did to the others in the family. While most of them responded, Atoulie held his ground. He was not going to let go so soon. He was angry, and his anger was valid, was all that Atoulie felt.

Atoulie could never have imagined that the letters wouldn't stop coming, even after Neibu was gone forever. Later, holding his martyred brother's last letter in his hand while waiting for his mortal remains to arrive, he couldn't concentrate on anything other than his last words, which read: 'Why don't you write to me? I am your loving brother.' Atoulie would read it over and over again, until his vision turned blurry due to the pools of tears gathering in his eyes. Eventually they would get too heavy and roll down his cheeks. He would wipe them off and read it again, harbouring the unimaginable pain and regret that seemed to be consuming him.

When Neibu's mortal remains were finally being buried, Atoulie looked at his brother's coffin for the last time, said a little prayer and told him silently, 'I am sorry, my brother; we could not understand one another.'

In the days following Neibu's ultimate sacrifice, as Atoulie found answers to all the lingering questions in his head, he also made another promise to his brother. He vowed to make sure that the story of his valiant and sincere brother would not die with him.

After he got commissioned, Neibu found the most pride in correcting people who wouldn't append his rank before his name. Perhaps nothing else gave him more joy than achieving that rank.

My brother's name is Captain Neikezhakuo Kengurüse, and this is his story.

1
Sense of Belonging

IN THE TAPESTRY OF life, I feel that there are those among us who bear the indelible scars of time, etched not just on the skin, but deep within the soul. Our survival is not a mere act of defiance against all odds, but a testament to the resilience of the human spirit, a beacon that guides future generations. For in our narratives lies the wisdom of ages, the lessons of history, and the unquenchable hope for a tomorrow that remembers, learns and grows from the yesterdays that we have lived and survived to tell others about.

I hope that I, Neingutoulie Kengurüse (you can call me Atoulie—pronounced as 'a-too-ley'), will do justice to the story I'm about to narrate. After all, it's a crucial one for my family, for my state Nagaland and, most importantly, for my dearest, wisest, elder brother, Neikezhakuo Kengurüse.

The real story of Nagaland is written in its hills, the magnificent festivals, its warriors and the everyday lives of its people—people that make this mountainous state what it is. Resilient. However, I blame the fortunate, or should I say unfortunate, location of Nagaland for its fate. Situated at the crossroads of the most scenic and unadulterated states on planet earth—with Arunachal Pradesh to its north-east, Manipur to the south, Assam to the west and north-west, and the country of Myanmar (Burma) to the east—Nagaland could be called the 'centre of attraction' for Northeast

India. Hence, it's no surprise that everyone would like to have a piece of the pie.

Nagaland is a supposedly small state, covering an area of 16,579 square kilometres, yet it consists of diverse tribes that have very different languages and cultures. There are seventeen major tribes—Angami, Ao, Chakhesang, Chang, Kachari, Khiamniungan, Konyak, Kuki, Lotha, Phom, Pochury, Rengma, Sangtam, Sümi, Tikhir, Yemchunger and Zeme-Liangmai (Zeliang)—all of which have hundreds of sub-tribes, which Nagas themselves sometimes find hard to remember. When I was a kid, the only language my village-bound family and I spoke was Angami, also known as Tenyiedie. Hindi and English, as languages, came to us much later as we started to communicate with the outside world for education, jobs, trade, etc.

Historically, our complicated relationship with India was a point of debate between the Nagas and mainland Indians. The fact that we supposedly don't look 'Indian', we don't speak or understand the official languages, we eat what we grow, and we live so far away that the national newspaper reaches us a day or two late doesn't help this case either. Yet, we were named the sixteenth state of India. And so, the debate continued.

For India, at large, the perception of the 'Northeast' is distorted by its geographical distance from the capital—New Delhi. Because of it, many people elsewhere in the country keep distance from the ongoings in these areas; it perhaps doesn't hit home like it does for us Nagas. Shaped by our histories of living close-knit, insular lives, we Northeasterners were, for a long time, treated as foreigners in other parts of the country. Thus, there used to be an 'anti-India' feeling among some of us, which sometimes got validated. One instance of this happening was when we heard rumours that a mediocre bureaucrat from another part of India is in Nagaland for a 'punishment posting'. It made you think—are we so intolerable?

For centuries, Nagas were able to convince rulers to leave us alone to determine our own fate. From the memorandum that was sent to the Simon Commission in 1929 to the formation of the Naga National Council (NNC) in 1946, it has been a long road to independence for us. A key landmark on this journey was the Nine-Point Agreement signed in June 1947. Among other things, this agreement gave us Nagas autonomy in relation to land-related matters.[1] However, this very pact became a bone of contention between Nagas and the government of India around the time of India's Independence Day. Finally, the NNC, which was under the leadership of Angami Zapu Phizo at the time, declared 14 August 1947 as the Naga Independence Day—a day before the Indian Independence Day. Therefore, a new era of conflict began for the so-called independent Nagaland.

You see, typically, Nagas are peaceful, god-fearing humans. However, despite the Naga Peace accord being signed in 2015, Nagaland has been conducting negotiations over another important document with India for years—the ceasefire agreement[2]—a piece of paper that the state feels will define its identity for future generations. Ironically, while the immediate goal of a ceasefire typically is to stop violence, this document has caused a stir among young Nagas across the country in recent times.

In Nagaland, we have an 80 per cent literacy rate today even though education and documenting history came to us as recently as the year 1960. Hence, not much is known about our 1,700-year history before we came under British rule. Therefore, it's understandable, to some extent, that more often than not it is difficult for common Indian citizens to pick a side in the Indo–Naga debate.

For Nagas, at the heart of the debate is their quest for a sense of belonging, a deep human yearning, akin to the roots of a tree seeking nourishment. It's not merely about being physically present in space but finding resonance in the intangible—in collective memory, cultural rituals and a silent understanding among people. In Nagaland, this quest is palpable in the way its local communities bond over folklore, shared struggles and collective triumphs—just some of the things Nagas don't have in common with mainland India.

It was this community that I was born into in 1976, much after the bloody conflict between the Nagas (mainly my community, the Angamis) and the Indian armed forces had eased out, yet the tension around the subject of 'independent Nagaland' could be felt in the air. The Shillong Accord had just been signed in 1975, unconditionally making the state a part of the Union of India.

Maybe this move by Shillong, the capital of Meghalaya, gave the Indian government hope that other Northeastern states would follow suit. However, some Nagas were determined to continue the fight for territorial independence.

None of us, whether you call us the current generation of Nagas or Indians, completely understand what the conflict is about. But history weaves a story rich with threads of aspirations, misunderstandings and broken dreams, reminding us hopefully that the path to reconciliation and peace is through embracing our common humanity and learning from the past to build a future where diversity is not just tolerated but celebrated.

The narrative for the Nagas changed a lot in post-Independence India.

On 14 August 1947, one day before India woke up with a true sense of purpose and freedom, the Naga National Council pressed ahead with its demand for a separate nation. Meanwhile, the Indian government had proposed that Nagaland become a crucial part of the Indian Union. It had been days of unrest for the Nagas as the British Indian forces began to pack their bags, concluding India's long-drawn freedom struggle, which had started back in 1942 with the Quit India movement.

Until 1947, the Nagas had fought in World War I and II for British India. The Battle of Kohima in 1944, a turning point in World War II and one of the deadliest battles the Japanese have ever participated in, was the first time when mainland Indian freedom fighters, who had come together as the Indian National Army (INA), along with the Japanese army, fought against the Nagas in a bloody, aggressive armed conflict. I believe this was the starting point of the downhill journey of the relationship between Nagas and mainland Indians. The Nagas were loyal to the British, and the Indians were committed to their fight against the British.

While some Indian patriots like Mohan Singh (founder of the Indian National Army) and Netaji Subhas Chandra Bose had found their life's purpose in a 'free India' by becoming allies in Japan's fight against the British army, the Nagas had found their peace in British rule itself. The British were happy with Naga warriors, and, as a reward, they diligently followed a policy of non-interference in Naga areas, hence earning the gratitude of the locals. Therefore, the Nagas wanted to continue with this arrangement if they could.

In their remoteness and isolation from the rest of India under the rule of the British, Nagaland had found a modicum of independence, so to say, and its villages thus had their own respective codes of behaviour and standards of administration.

Now, in 1947, amidst the Partition of India and the rearrangement of states as well as citizens of the world, India wanted Nagaland to be the sixteenth state of the country. Instead, the newly formed NNC requested the Indian government, led by then Prime Minister Jawaharlal Nehru, to let Nagaland remain independent and not submit to the Indian Constitution. Nehru responded diplomatically to the tricky ask by saying, 'Our policy has always been to give the fullest autonomy and opportunity of self-development to the Naga people without interfering in any way in their internal affairs or way of life. The Nagas are as independent as any other Indian citizens.'[3] The statement could have many interpretations.

On the other hand, the father of the Indian nation, Mahatma Gandhi, had pacified Nagas perplexed by the idea of being a part of India. He had said, 'Personally, I feel that you all [Nagas] belong to one [side], to India. But if you say you won't [come with India], no one can force you.'[4] This statement too could have many interpretations.

That's the thing about words—if taken out of context, they can lead to families breaking, governments dissolving and wars breaking out. Mahatma Gandhi's words did calm down the Nagas, but they gave them hope of an independent nation—a promise that could not be delivered.

Days and years passed by and the Nagas who were clueless about what had transpired, especially our grandparents and great-grandparents, waited for their identity and a sense of belonging in India—until they felt that they had to take matters into their own hands.

Finally, a plebiscite organized on 16 May 1951 by the Nagas internally revealed that over 99 per cent of the voters upheld full Naga sovereignty. In December of the same year, the then NNC president Angami Zapu Phizo, who had first declared independence for the Nagas in 1947, told Nehru about the plebiscite, who rejected the proposal. Resultantly, no vote was cast in Nagaland during the first Indian general election of 1952, since 'Indian' was not the identity the Nagas were looking for or wanted to associate themselves with. People from our community are sometimes called 'Chinese' when they visit or migrate to any Indian state that is not in the Northeastern region.

In September 1954, Phizo—today, famously known as the father of the Naga nation—broke his silence and formed the Free Naga Government. Later, he wrote a letter—this time presenting his views to the first President of India, Dr Rajendra Prasad—asserting the fact that the Nagas had a 1,700-year history as an independent nation. In his words: 'We do not want anything from India, please leave us alone.'[5]

The Indian government, which had initially overlooked the declaration of independence, responded in the face of greater assertion of Naga nationalism by arresting nationalist leaders and raiding their villages. As a response, in December 1955, Phizo announced plans to establish an army to defend Nagaland. A violent insurgency soon ensued.

A few Nagas had the confidence to take on this battle since even though the British had departed from Nagaland voluntarily in 1947, they had left us with two crucial practices that instilled resilience and strategic prowess in our community—Christianity and modern warfare with weapons. Christianity, of course, is what brings us together and is prevalent in many aspects of our everyday lives, and this should become evident later in this book, but for now, let me emphasize the latter point.

Until World War II, even when the Nagas fought against the Japanese forces in the deadly battle of Kohima, they stuck to their traditional war practices, such as headhunting. As hard as it may be to imagine today, in headhunting, a man kills his enemy, cuts off his victim's head and carries it home as a trophy. Our great-great-grandfather, Perheile, fought in that war, and he was the greatest and most famous headhunter of his times.

So, equipped with learnings from the time of British rule, the Nagas firmly believed that they were ready to take on the world for their independence. But, as we have witnessed, that was not to be.

The 1950s marked an era of profound turmoil for the community, characterized by severe disruptions to rural livelihoods, deep-rooted bitterness and stark divisions. Tragically, this was merely the beginning of Nagaland's descent into chaos.

In the period between 1954 and 1956, the Naga National Council declared that at least one out of ten members in any household would have to represent their family name in the underground Naga Federal Army (NFA), and the other nine would take care of the tenth family member whenever he or his family needed it while they were at war. From the Kengurüse family, my father's elder brother, Lhounolie, represented us. Like a typical traditional Naga family, my father, Neiselie, had a large family—with five brothers and four sisters—but by the time I was born, I only saw three of my aunts and two of my uncles. While Lhounolie joined the Naga army, the rest of his siblings took to cultivation to support the family.

The year 1956 particularly was that of rebellion for Nagaland. Those whom some call our freedom fighters had decided that enough is enough. Some of us Nagas had picked up arms, stopped paying taxes and extremists continued to sabotage the Indian Army. In June of 1956, just as the Naga army attacked a group of Indian soldiers in Kohima, the Indian Army called for reinforcements and started to retaliate. The cycle of violence in the small empire of Nagaland thus began, and it led to a series of destructions in Naga villages and of its people's lives.

Visier Meyasetsu Sanyu, a Naga historian, presents some of his memories and the regrettable circumstances from back then in his book, *A Naga Odyssey*.[6] It paints a heart-wrenching picture of Nagaland from a time when its peace was disrupted, almost overnight, as the Indian armed forces launched a counterattack. Stories like this one deeply impacted many Naga children like Sanyu who were born at the time, including my father, mother, uncles and aunts. Sanyu's story is from the village of Khonoma, which consists of three *khels* (clans)—Merhüma, Semoma and Thevoma.

Picture this.

On a night in June 1956, our family of nine—my mother and father, three brothers and three sisters and I—slipped into the darkness. Most members of the third *khel*, Thevoma, who supported autonomy within India, stayed there under the protection of the Indian Army, but a few joined us. Indeed, some Thevoma joined the village guard set up by the Indian military and fought against their own people. This was a time of great village upheaval, bitterness and division.

My father took his gun—loaded of course—for we were in great fear of discovery. Others carried spears and axes. Most of us lugged baskets brimming with food and essential clothing, and any precious items such as jewellery. My mother took coins loaded in a belt. We all wore our Angami tribal shawls about our shoulders to keep us warm. As I was just five, my eldest brother Perhicha, then in his twenties, carried me wrapped in a Naga shawl slung over his back like a rucksack. We left the village walking past our rice fields in the dark, then we climbed and climbed to put some distance between us and Khonoma.

I dozed in my sling, bumping against my brother's back. A sharp pain caused me to cry out as a nettle bush scraped my legs, causing one of my family to gag my mouth to avoid detection.

For all of us it was an emotional time, although being five I had no real sense of what was happening. Yet now that I understand, I become emotional just remembering this leaving of home.

Since then, separation from my home has been a constant feature of my life.[7]

Since the onset of events like the one cited above, Nagas who witnessed these atrocities in the 1950s and beyond have not been able to overcome their fear of armed conflicts to date. This includes some of my own family, like my aunt, Medolhouü Rio. She often talks about her earliest memories—from the time just after World War II as she was born right in the middle of it in 1945. She and her family had witnessed the 1955–56 Indo–Naga conflict closely, where the Nagas were kept in groups by the army and they put 'pointed things' on the road to stop them from moving around, even preventing them from fetching water sometimes. She remembers that all the houses in their neighbourhood were burnt, and locals were tortured in rooms.

It was a time when even using the roads in Nagaland was scary.

When the government of India could no longer curb the trouble brewing in the Northeast, they gave special powers, under the Armed Forces Special Powers Act or AFSPA, to the Indian Army in 1958 in the states of Arunachal Pradesh, Assam,

Manipur, Meghalaya, Mizoram, Nagaland and Tripura. Instead of just maintaining law, order and peace, as the army would do during peacetime, this Act allows the Indian Army to gain the power to take any action against locals as long as it can justify it as an attempt to maintain order in 'disturbed' areas; in other words, it can do this without having to face the usual legal consequences. The cycle of violence in the small empire of Nagaland, it seems, was triggered around this time.[8]

Until then, Nagaland was still a district in Assam then known as 'Naga Hills'. Although the armed insurgency helped Nagaland earn its own identity as a state (independent of Assam) by 1963, its own Constitution and flag were still big question marks.

Parallelly, the Indian government had rejected Phizo's demand for Nagaland to be allowed to function like an independent state. Despite the various amendments and agreements on the ceasefire agreement over the past seventy-plus years, peace has not arrived, it seems, in Nagaland.

My first impression of Nagaland, in the 1970s, was extremely different and in stark contrast with what my aunt Medolhouü had faced.

The time my brothers and I were born was peaceful yet rocky on the personal front—it was the 1970s, and the armed conflicts had fizzled out a bit. The restlessness in the air, however, remained.

In 1976, I was six months old, my brother Neikezhakuo (Neibu) was three years old and our eldest brother, Rüükielie, was about six.

Our newly single father could not take care of three young boys all by himself while fulfilling his role as the sole breadwinner of the family. So, my brothers and I were sent from Kohima, the picturesque capital city of Nagaland, situated in the south at an altitude of 1,444 metres above sea level, to Nerhema village, to our parental grandparents, Thepfulhousa and Thenuzieu.

Back in the village, we spent most of our time at home, waiting for our grandparents to return from the paddy fields and feed us. We would cry, *'Apuotsa, mhai vorlie cie … merü thor bate ho …'* (Grandpa please come back home … We are very hungry.) My brothers and I were blind to the fact that there was a civil war going on in the state. All we knew was that a new 'normal' in our life was beginning. We were only consumed by the questions in our immediate life at that point: Why has *Apfü* (mother) packed her bags and left home? Why are we not allowed to see her? Why are we living with our grandparents? Why are we sent to a school made of thatch? Why can't we afford one toffee for the three of us?

Overnight, life had changed for us three brothers, and our silent suffering began the day we realized that our parents were separated. Our family was split into two while we were already dealing with our poverty (owing to the poor Naga economy and lack of opportunities in the post-conflict times), and we were left with no say in the matter. But, as they say, good things happen to good people; so, this dark time for our father and us brothers ended in a few years.

Very soon, our family of four became a family of thirteen, as our stepmother, Dinuo, gave birth to ten children—Kikrusenuo, Khrisaneisa, Keneitsilie, Neiketuü, Seyievituo, Arhenuo,

Nimbu Saab

Mhozienuo, Thepfulie, Neichünuo and Viketoulie. The eldest siblings, in this case, were Rüükielie, Neibu and me—in that order. There are so many of us today that sometimes, I forget the names of my own brothers and sisters while talking about our elaborate family tree.

My birth mother, Nupulhuu, on the other hand, also gave birth to four kids after us with her second husband, Dunucho Swuro—Banuo, Zebulie, Khotuo and Kenienuo.

As the family became bigger and happier, my brothers and I were drawn closer and closer together. But one question remained for us as it remains for Nagaland still—where do we come from, and where do we belong?

2
God's Plan

LET ME TAKE YOU back to our journey from the very beginning. Perhaps, the one peculiar difference in our culture in Nagaland that could be jarring to the people of the other states, especially those that are not in the Northeast, is that many marriages do not involve any formal ceremonies. It is common for a boy and a girl to fall in love with each other and, with mutual consent and love, start living together as a married couple. Perhaps it would not be wrong to say that most marriages here are a result of love, and not really the family or society's matchmaking efforts.

My birth parents were no different than most of the young couples in Nagaland. My father, Neiselie, and my birth mother, Nupulhuu, were cousins. It wasn't unusual in our state for two people of the same family to get married to each other. There were many marriages that had taken place due to such a mutual understanding between a boy and a girl who belong to the same extended family.

My father, in the early '70s, used to run a small grocery store in Kohima. My mother, who knew him from his visits to her house, would go to his shop to get groceries for her family regularly. Soon, the short exchanges turned into longer conversations, and the young couple fell in love.

Love meant living together, which later translated to marriage. Within the next few years, Neiselie and Nupulhuu became parents to three young boys: Rüükielie, Neikezhakuo and me, Neingutoulie.

There's a story behind our names, as a lot of thought went into them. When my father had asked my mother for marriage, she had readily agreed. So, his first son's name Rüükielie meant 'to get easily'. When his second son was born, he thought having a second son was going to give him much more joy; hence he named him Neikezhakuo, which meant 'greater happiness'. When my mother was pregnant the third time, my father thought, how nice it would be if he could be the father of three sons. So, he named me Neingutoulie, which meant 'exactly what was desired'.

However, unfortunately, soon after my birth, my parents started having differences with each other. They developed trust issues that led to misunderstandings. Soon, it became a cycle. For example, if my mother were delayed in getting back home, my father would raise questions about her whereabouts. The more questions he raised, the more my mother rebelled and would spend even more time out. The last straw was the day my mother came home late from her own mother's place and my father asked her, 'Who are you having an affair with? Do you want to marry him?' Tired of the unending accusations, my mother packed her bags and went back to her parents.

In the beginning, my father asked her to get back home since they had three children together. But my mother, who was young and hurt by the fact that she had been wrongly accused, refused. However, after some days of separation, when she started missing us and came with her mother to take us away with her, my father and grandfather refused to let her do it. In our state, if and when a couple separates, the sons usually go with the father, while the daughters go with the mother. Given these traditional beliefs, they were not ready to entertain any discussions about who we were going to live with. My mother begged and pleaded with them, since Neibu was not even four years old at the time, and I was barely six months old and still being breastfed by her. But their decision had already been made. A firm no was pronounced, and my mother returned with only me in her sore arms, feeling helpless, and asked my father and paternal grandparents to take care of us. However, I too was taken away from her soon after, and the three of us were cared for by our father thereafter.

While we spent a few days with our father in Kohima, after their separation, he eventually realized that we were too young and that, without a loving mother at home, it wasn't possible for him to look after us while taking care of his small business. So, he decided to take us to our grandparents, who lived in a small village called Nerhema, further upwards of Kohima, about 23 kilometres via road.

We stayed in Nerhema for three to four years. Our days were spent playing and exploring the small but beautiful little hill where our grandparents' house was situated. I was no more than a baby, and so, while my brothers played and ventured out, I would just follow them around. Our favourite time of the day used to be when all of us would gather around the fireplace to have our meal. One of my fondest memories from this period is that of Neibu and Rüükielie teasing me about my large appetite despite being only a child. I would often finish the portion of meat served to me, which was smaller than everyone else's since I was the youngest, and I would invariably have to walk up to my grandfather with my plate to ask if I could take some extra meat from his portion.

Though we were away from our parents, my brothers and I didn't completely miss the warmth of paternal love, owing to the care of our grandparents. But our father missed us, and so he would come over to the village every weekend. We would rush to him, always searching his pockets for sweets that we knew he would carry for us from Kohima. In those times, sweets were a rare delicacy, and we truly savoured them.

Our lives were very simple, and it wouldn't be wrong to say that we were poor. For instance, I remember, one time a uniformed inspector had come to inspect the thatched school that my brothers and I went to in the village. Before returning, he distributed some sweets to the younger classes, including mine. It was a small boiled, orange-flavoured confectionery, only slightly bigger than a pea. All of us students belonged to poor families, so receiving an unexpected surprise like that seemed no less than a feast for us. However, I realized that it was only my class that had received them, and my brothers had been left off. So, while everyone popped their candies almost the minute they received it, I saved mine to share with my brothers.

As soon as I saw them during our recess, I flashed my prized possession at them. All three of us got excited. I could now finally unwrap and taste it. I put the small round ball between my teeth in order to snap it into three pieces and enjoy it with both of them. To my horror, the minute I bit into it, it didn't snap and instead slipped right through my throat. I had unintentionally swallowed the whole sweet! Not only were my brothers denied the joy of having it, I, too, couldn't even taste it. The disappointment of losing out on the feast that we had been looking forward to so much was written large on all our faces. We all stood dumbfounded for a few seconds; the awkwardness of the moment was finally broken by Rüükielie stomping off with an annoyed shrug.

I kept feeling regret at spoiling what was supposed to be a moment of great joy for the three of us. Hunched, I turned to look at Neibu. He too was utterly saddened but simply looked at my face kindly and said, 'Aye Atoulie, you should have shared it with us.'

When school got over, by the time I could spot my brothers, I realized Rüükielie had left and it was only Neibu who was standing outside the school, waiting for me by himself. As I reached near him, he held out his hand to take my bag; with the free hand, he swung it on his free shoulder, and both of us quietly walked back home, still visibly upset at losing out on a little moment of joy.

Unfortunately, I didn't remember our mother at all. If my brothers did, they never spoke of her to me, or in front of me, during our days in Nerhema. For us our grandparents were the parents that we knew, and our father was our special weekend visitor who always came bearing gifts. We didn't miss anything or anyone at the time. It was a simple, but happy, life that we lived.

One day, our father came, for one of his weekend visits, with another woman. At first, we didn't even notice her; as usual, we leapt towards our father's pockets to look for sweets first. But this time, his pockets were empty. As we retreated from him, a little taken aback, we realized that he wasn't alone. There was another person in the room—the woman we didn't recognize.

When they came in and sat with our grandparents, the three of us went to a corner and kept looking at them. We were quite amused, since we had not seen our father with another lady until then. Neibu and Rüükielie kept looking at them, saying, '*Daddy ladki leke aaya hai!*' (Daddy has brought home a girl!), and they were laughing about it. Not wanting to be left behind, I joined them, too. So, for the first few minutes, the three of us kept making jokes about it.

The elders were talking amongst themselves while my brothers and I listened. We gathered that the lady's name was Dinuo. She was my father's neighbour in the town where he lived and had taken very good care of him when he had got hurt in a small accident. During this time, they had fallen in love. Her aunt had apprised her of the fact that our father had three sons, and she had decided to be with my father regardless. My father, who was lonely and needed a partner to be able to bring us up if he wanted us to be with him, couldn't be happier with her decision and decided to bring her to Nerhema to meet us all.

We felt their gaze shift and focus on the three of us standing in the corner. Since I was the youngest, the lady with my father started beckoning me to come meet her. Fascinated, even though a bit confused, I walked up to her, and she made me sit in her lap. My ears turned red with embarrassment, since I could see my brothers in the corner and knew that I too would become the butt of their jokes now. They were pointing at me and having a hearty laugh, making the colour of my face turn to a light shade of red.

God's Plan

Eventually, we were told that she was our new mother. Since I had no memory of my birth mother, unlike my brothers, this news did not evoke any emotion in me. However, they seemed to accept the new normal politely, and so I followed suit. We started addressing her as Apfü.

By this time, my father had also left his work in Kohima and had been able to secure a government job as a surveillance worker in another small town known as Jalukie. Since, as a new couple, it was a fresh start for them, my grandparents suggested that they go to Jalukie alone, assuring them that they would take care of us in Nerhema. Dinuo turned down this offer, saying we were her children too. My father too did not want to separate us from them. Hence, it was decided that they would first take me and, eventually, the other two would join us and settle down in Jalukie.

Once again life changed for us. I had never lived without my brothers. But I had to follow what my father said and went with them to Jalukie. In Jalukie, since my father would be away for work, I spent most of my time with my stepmother. She would feed me and play with me the whole day. I liked her. I started enjoying my time with her. However, I still missed my brothers a lot, especially Neibu, with whom I was closer. It wasn't the same without them.

Thus, after a year, when my father finally brought both my brothers to Jalukie, I was very happy. We could finally be a family just like everyone else, I felt.

Jalukie was where our father transformed from an occasional weekend visitor into the head of the family for us. Back in Nerhema, since he would only be with us for a day or two, we only got to see the fun, indulgent side of him, but in Jalukie we met the hard taskmaster who did not entertain any excuses when it came to some things, especially studies. All three of us were in school, and we were told clearly that should we fail our exams, no Christmas gifts were to be expected from our parents. He even implemented the decision when Neibu, who had just moved schools and was thus unable to cope in the first set of exams he took, failed. The punishment that had been laid out was duly carried out.

However, reuniting with my brothers was a blessing for me. Because Rüükielie was a bit older and so did not involve the younger ones much socially, Neibu and I were joined at the hip to each other. Whether it was school or the church, we went together everywhere. Sweet as he was, he would often even carry my bag for me. Sometimes we would simply follow Rüükielie and watch him as he would go play with the older boys. Even though he could have sometimes, Neibu never left me alone. I would always feel safe and protected with him around.

In the first few years, the three of us would walk to school together. On our way back, since I was in a lower class, my classes would finish earlier. But I used to sit and wait for my brothers, and especially Neibu, to finish for us to be able to go back home together. Rüükielie already had a lot of friends, so he wouldn't always be with us on the way back.

One time, while we were on our way back home together, we saw a bird nest at the top of a tree. When I pointed it out, Rüükielie said, 'Is it a nest? Let's take it out.' We were curious to see what would be inside the nest since we had never seen one from the inside before. Thus, Rüükielie, who was the tallest and strongest of us all, quickly climbed up. Watching him, Neibu also followed and went up the branch below Rüükielie's. I stood on the ground, looking up at both my brothers roughing it out, oblivious to the big muddy patches that each brush of their limbs with the dusty bark of the tree was leaving on their clothes. The dirt on their clothes aside, all the hanging they had done from the branches had caused their pants to begin to slide off their bony waists. Especially Rüükielie's, since he had climbed further up and hadn't had a chance to pull them up. Lo and behold, just as he was about to reach the nest, the waistband of his pants gave way, and as a result, his pants went swaying down like a bird. What made matters worse was that, in those days, we didn't always wear underpants, and so Rüükielie, who hadn't worn any on that day, was left dangling with nothing on the lower half of his body—seeming almost naked. I tried my best not to react, but one look at Neibu made us both burst out into guffaws of unstoppable laughter, until our eyes teared up.

Turning almost red with embarrassment as well as anger at us for laughing at his misery, Rüükielie looked at us and said, 'This is what you do instead of helping your brother? Is this how you behave if you love me? Let it be, I will not carry any nest down. I will come down now.' Not wanting to embarrass him any further, and desperate to see the contents of the nest, we quickly tried to backtrack, contain our laughter and plead with him, begging him to not get angry and to bring the nest down. We were not going to lose an opportunity to look inside the nest. Placated, Rüükielie proceeded to carry it down carefully; however, it seemed the bird had abandoned it. We could not see anything other than some small black round bits inside. Neibu told me that it seemed to him that the nest belonged to a wild sparrow. Our little adventure ended rather disappointingly, and we had to walk quietly for the rest of the way behind Rüükielie who was still upset with us for having laughed at him when he was stuck. Neibu and I were careful to not look at each other, as that would have made us break out in laughter again, and so the walk back home turned out to be another challenge.

A big part of our lives, which holds its importance till date, is the church. In fact, our father was a pastor in the church in Jalukie, so Sunday church was never missed. Out of all of us, it was Neibu who was the sincerest towards God and our religion. Having our father as the pastor also meant that we were brought up with strong values. He was as strict about them as he was about our studies, if not more.

We knew and understood it well too. Therefore, when Neibu and I happened to find a wallet one time on our walk back from school, we were more scared than happy. Upon finding it, we tried to check to see if there was anyone around, but the road was empty and there was no other way of identifying the owner since it had nothing in it but money. At first, we decided to give it to our father, but we both agreed that he might not believe that we found a wallet full of money just lying on the ground. Hence, we both agreed to spend it quietly, without telling anyone. For a day or two we would just buy some of our favourite things to eat from the vendors on the roads and finish it before reaching home. However, Rüükielie noticed us doing it and confronted us about it. We told him our story and even offered him ₹10 to keep our secret. He accepted it and agreed to keep it to himself. However, once the money was over, he promptly went to our father and told him. Our father, furious at us for hiding the truth before and lying to him, denied us dinner that night as punishment. That taught us to never even think of being dishonest.

In the next few years, as we grew older, the size of our family grew too. Our religion didn't allow practices like family planning, as it was believed that that was going against the way of God. This meant that for a long time our family grew larger and larger. Slowly the pressure of having one earning member but many mouths to feed started becoming a concern.

This is when the differences in my brothers' personalities started becoming clearer. While Rüükielie was the oldest and also the smarter one, he was more outgoing and seemed to usually be busy with his friends and other activities outside of the house. Neibu, on the other hand, seemed to treat all his younger siblings with more responsibility even as he treated the elder one with respect. Rüükielie and Neibu's friends would often wonder how even though they were brothers, they never seemed to fight amongst themselves. Neibu always treated Rüükielie with respect and stood a step behind literally and figuratively. His focus remained not only on his but on his siblings' studies, as he became cognizant of the fact that it was only education that could uplift him and others in the family.

At home too, when we started getting more siblings, I would sometimes feel that our stepmother, whom I was the closest to since I was the only one who had spent time with her alone in the first year in Jalukie, had started changing towards us. Maybe she cared more for the children that came after. More often than not, it would be Neibu who would be the voice of reason for me. He would tell me how I was most likely misunderstanding it, reassuring me that she cared more for the younger ones only because they needed more care due to their age. While I started drifting away from her, Neibu continued to remain the dutiful son who always listened to our parents, making sure he did everything that was asked of him.

Neibu had perhaps started maturing beyond his years very early. I remember when he was barely in the seventh or eighth standard, he had to give a speech in assembly at our school, and the topic he had chosen was 'fair-weather friends'. Time has blurred my memory about the contents of the speech he gave, but one line I never forgot was, 'A fair-weather friend is like a shoe with a hole in it. You don't spot it until it is raining.' I always wondered where this wisdom came from.

In the years to come, we became a family of thirteen siblings in total. We spent the majority of our time in Jalukie, busy with school on weekdays and in church on weekends. Neibu especially really believed in and enjoyed church work and would enthusiastically take part in all church activities—singing and even teaching others who couldn't sing. The rest of the time—vacations and other holidays—were spent in Nerhema with our grandparents where we would give them a hand in their paddy fields when not spending time playing with our friends.

We may not have had much to spare, but we also didn't have a lot to complain about, for we all believed firmly in God's plan.

3
Men Don't Cry

THE INSURGENT ACTIVITIES AND political unrest in Nagaland, from as early as the 1950s all the way until the early 1990s, were like distant thunder—ominous but not always directly impacting our daily lives. As Nagas, we were still living under the fear of men in uniform, and random checks as well as shootings were still very common in some areas. However, the changing perception of locals about the Indian Army and the political ongoings did indirectly affect us. Discussions at home sometimes revolved around these issues, although our father's role as a pastor and his apolitical stance shielded us from getting too involved in these matters.

Unfortunately, this also meant that there were not many conversations between my brothers and me when we really needed them. You see, the one thing men across the world have in common is that we're told since we are little to 'Be a man', or that 'Men don't cry'. Since childhood, my brothers and I were good at following rules, so we just obeyed—we didn't cry in front of each other or even alone. In Jalukie, we simply went about our days, from one responsibility within the family to another, as men in most societies would, without ever addressing our internal struggles.

We were missing a crucial piece of the puzzle in our lives—an Apfü-sized hole in our hearts. Although we had a new Apfü in our lives, life would never be the same again for us three brothers. So, we tried our best to go on with our lives as if nothing had happened in the family and everything was normal. The fact, however, was that nothing in our lives was normal anymore; everything had happened so suddenly. On most days, we succeeded in masking our feelings, but on some days, we failed and broke down.

For us, shifting from Nerhema to Jalukie wasn't just a move from a village to a town; it was also the advent of a new family structure, which came with its own complexities and challenges for us brothers. And despite the presence of our new Apfü (our stepmother), there were times when the emotional gap was palpable. The family felt incomplete to us. We often pondered over our scenario in our heads—the rapid shift from our birth Apfü to our step Apfü, an ever-increasing number of siblings, and our father's struggle to make ends meet with his new occupation. These reflections sometimes led to existential questions that clouded our minds. As we felt more and more distant from our father's new family, all three of us felt the loss of our mother more deeply and gravely, but each of us learnt to cope with things in our own ways.

Neibu devoted himself to the service of God and academia, the two things that could, perhaps, increase the probability of good luck arriving and change the fate of our family. Rüükielie, on the other hand, the smart young gun that he was, didn't seem to care about these things too much. Besides, he didn't have to work hard to do well. So, he spent most of his time outside the house, with friends, always thinking of ways to move out of Jalukie. And I, the youngest one out of the three of us, turned towards 'finding humour in tragedy'—something that Neibu disliked because he saw my mischief as carelessness. He would have ideally liked me to study more and focus on getting good grades at school, while I would just study when I felt like. I would roam around and go for movies when I felt like it. The way Neibu worked hard in life, his willpower and his grit were inimitable. So, unlike what happens in many Indian households, with kids being compared to their cousins, in my house, I was compared to my elder brother all the time—'*Aap toh Neibu jaisa nahi hai, aap toh aisa hai* (You are not like Neibu; you have your own ways),' they'd tell me.

After a point, I stopped caring and started acknowledging the fact that all three of us brothers had been gifted with three contrasting personalities by God, making us a perfect trio that complemented each other and helped us to get through life.

When I look back now, it does dawn upon me that perhaps people were right—there was no one like Neibu. I could never be like Neibu even if I wanted to. His determination to change our fate, which he developed before we could even recognize our unfortunate socio-economic reality, was beyond me. Burning the midnight oil before his class ten exams, he would sit wrapped up in a square blanket, with his books. He wanted to score good marks in order to make it to a well-known science college in Kohima. Even if he wanted to take a nap while he was pulling off all-nighters at his table, he would lean over and rest his head on the arm of the chair and then wake up again to study after the nap. To get ahead in his studies, he would subscribe to all kinds of English-language newspapers, including *The Hindu* and *The Times of India*, for ₹30 every month.

At that point, people in Nagaland, like in many remote areas of India, would get all Indian newspapers three or four days late due to a combination of geographical, infrastructural and logistical challenges. The political unrest and insurgency didn't help either, they would regularly disrupt delivery and distribution channels. To add to Neibu's misery, I would pull his leg by telling him, 'Neibu, this is not news; this is stale news.' That was the kind of distance then, literally and metaphorically, between Nagaland and mainland India.

Neibu's dedication to academics and the church was unparalleled, it kept him busy; yet there was always a fire within him that yearned for something more, something beyond the boundaries of our village and the small world we knew. I sometimes craved for the focus he had, which he directed towards securing a bright future for our family. He firmly believed we could get out of our poverty-stricken situation; his optimism not only seemed endless then, but also enviable.

One other thing that I wish I could have learned from Neibu was his maturity when it came to handling different relationships. At home, with our parents, he was a different person, and with the siblings, he was like a parent—completely different. Similarly, outside the house, with friends, he was the most approachable and fun-loving human.

I was sometimes awestruck by the relationship Neibu had built with our parents. They were best friends, and he used to share all his stories from school with them, sitting with them until midnight, with the aim of making them laugh. When he was with our parents and in a good mood, the whole family, including the siblings, would gather around in the big kitchen. All of us would sit around him on *moodhas* (stools) during his storytelling time, laughing, sharing a meal and enjoying his anecdotes. Mostly, Neibu's stories consisted of jokes that made us all erupt in laughter, which echoed through our humble household. He had a unique way of hiding some of life's wisest lessons within his stories and jokes, especially when our youngest siblings were listening. Below is one of the jokes that stayed with me for a long time—a take on the power of manifestation, the art of simply asking the universe to make things happen for you:

In a remote and lush forest, there once lived a man with an insatiable appetite for adventure and the sweetest fruits. One sunny day, driven by his craving, he embarked on a daring ascent, up a towering tree. His eyes were fixed on a particularly succulent fruit, which seemed to beckon him from afar. As he inched closer, the branch beneath him, fragile but unyielding until now, snapped with a thunderous crack, sending him plummeting towards the unforgiving ground below.

In that heart-stopping moment, while suspended mid-air, the man cried out desperately to the heavens, 'Oh God, save me!' Almost as if in response to his plea, a miraculous thing occurred. The hand of fate intervened as the broken branch, guided by some divine force, caught him by the collar of his shirt, leaving him dangling precariously between the earth below and the sky above.

There he hung, a spectacle for the world to see, unable to climb up to his prize or descend to safety. In a moment of cheeky bravado, he called out again, 'Oh God, I was just joking!' But the universe has a peculiar sense of humour. As if responding to his challenge, the branch gave way once more, sending him crashing to the ground with a comical thud. Landing squarely on his buttocks, he exclaimed in a mix of pain and revelation, 'Oh God, you truly don't know how to joke!'

These light-hearted moments with our family often made us forget for a moment the financial situation at home.

Another activity that enabled us to bond together as a family was the church. Owing to Apuo's position as a pastor, and the fact that we were all living in the church compound, we attended the church service every Sunday, but Neibu would go a step further. He participated in youth service on Saturday evenings too. This service refers to religious gatherings targeting the younger members of church as it was tailored to address their spiritual, social and emotional needs, especially those of teenagers. Unlike Rüükielie, who had always been the more extroverted and seemingly carefree one of the three of us, Neibu was growing into a man of deep convictions, responsibilities, and great discipline.

The relationship Neibu and I had with our younger siblings was a blend of care and responsibility. We felt a sense of duty towards them as their eldest brothers. More importantly, we, as older brothers, were conscious of the fact that we were their caregivers, their role models and, at times, their disciplinarians.

This familial bond, while strong, also carried a weight of responsibility, which often felt overwhelming for our young shoulders. However, despite the risk of not being liked by the siblings, Neibu took on this challenge like no one else in the house—not even our parents. At a very young age, he had become a taskmaster to all the siblings, who were more scared of Neibu than our parents. None of the younger ones dared to make eye contact with him when they had done something wrong since Neibu's big, watchful eyes could pierce through any facade, discerning unspoken truths and masked unwarranted actions. He was a silent guardian and a model of integrity in every interaction he had with the children of the house.

In his role as the family's disciplinarian, Neibu adopted a firm yet fair approach, reminiscent of a drill instructor. He was vigilant and ensured that the household's rules were respected—a task he undertook with a keen sense of responsibility. There was this one time when two of our younger brothers, Keneitsilie and Seyievituo, were not in their respective beds on time and were nowhere to be found. When Neibu found their beds empty, he guessed where they may be. Because our family could not afford a television then, the kids of our house had come up with a way, or if I may say so as Indians do, a *jugaad* way, of entertaining themselves. As he had suspected, he found them sneakily watching TV from behind a wall, peering in our Bihari neighbours' house. The neighbours probably didn't mind it, but Neibu was concerned when he found out about the disobedient act. He patiently waited for them to finish watching what they were, and as they turned around, he caught them red-handed.

Neibu's problem was not the fact that our young brothers were watching TV; it was that they never asked any elder in the house for permission to watch it.

To make them pay for their mischief, he made Keneitsilie, who was the older one and in class five, stand next to the bed all night with his hands up, and since Seyievituo was a little younger, he gave him a more lenient punishment—standing without his hands up. Neibu didn't only give the punishment, he made sure that the brothers completed the task, as he sat next to them, monitoring their every move, without uttering a word all night. Perhaps, this was Neibu's way of teaching Keneitsilie and Seyievituo the importance of respecting and obeying the elders of the house.

In a hilarious turn of events, after a long, tiring and painful night, when Keneitsilie woke up the next morning and tried to wash his face, the water was splashing from right above his head, yet not a drop reached his face. His hands had given way. A hard lesson had been learnt.

During this time, Neibu was also growing into a charming man and a leader. Fortunately, his newfound love for academia and the church in Jalukie sparked some lifelong friendships. For instance, the service at church on Saturdays led him to his first crush, Akhrenuo. She was in class seven when Neibu was in the ninth. She ticked all the boxes of Neibu's idea of a perfect girl—she was a well-read, caring, sincere, religious, simple Naga girl with a calming presence. Akhrenuo had moved to Jalukie in class four, with her family, while we were already living there. Neibu was a couple of years older than her, and they went to the Baptist church together. Eventually, she also joined him for the youth fellowship there over weekends.

By the time Neibu was in class seven, we had moved from the government school where he had been enrolled until now to

St. Xavier's, a Catholic school in town. This was the time when Neibu became the centre of attraction for many girls. Apart from the fact that he had many more girls in his class, there was the fact that he was a caring and empathetic boy—a quality that seemed to draw the attention of every girl in his batch. However, he had eyes only for Akhrenuo. He even wrote to her about his feelings but, considering they were very young, she turned down his proposal, although she was very fond of him. Neibu didn't walk out of the situation with love, but he certainly earned a friend for life. Since both Neibu and Akhrenuo were preparing to enter the Kohima Science College and join the medical stream thereafter, this friendship proved to be very beneficial for both of them.

By this point, Neibu's leadership skills were also starting to shine bright—wherever he went. At church, he was the choir leader and the general secretary in the youth fellowship wing for a long time. In school, he had been appointed as the first-ever president of the Angami Students' Union. His fun-loving attitude and the spark in his eyes, which he directed towards everything and everyone, could always light up a dull room. During this time in his life, Neibu could have easily passed off as an evangelist, a pastor or even a priest. Some of my friends, cousins and I were certain he would get into politics, while others had their votes on him becoming a pastor someday, considering that his extraordinary public-speaking skill married well with his devotion to religion.

Meanwhile, in 1989, Rüükielie left home for Kohima after his matriculation, to pursue a commerce degree offered at Baptist College and newer adventures in the 'real' world, far away from the adversities back home. Neibu and I continued to stay in Jalukie and tried to make the most out of what life was throwing at us every moment. This move by Rüükielie created even more distance between him and us as brothers. He had his own life, habits and preferences, which we didn't understand or get into so as to give him enough space to deal with what was going on in our family. As the eldest brother, he also had to shoulder the weight of financial responsibilities he didn't sign up for. These thoughts began to consume him and, eventually, he shut himself off to not only the family but us too. While he was studying in Kohima, he hardly came back home. Even if he did, he'd come only for a day or two and leave as soon as possible.

Parallel to that, in Jalukie, Neibu and I were making new friends. Akhrenuo, Sanuo and Ravannuo were three girls who had a huge role to play in our lives, gifting us with many happy childhood memories, a rare occurrence for us during that time. I had become an official plus-one for Neibu at all his friendly gatherings because, unlike Rüükielie's friends, I could connect with Neibu's friends on account of them being closer to my age. However, while Neibu had more friends who were girls, I had more male friends. I had even befriended the younger brothers of Akhrenuo and Sanuo. Hence, this overlap of friendships resulted in us being part of one large group of friends that celebrated all festivals together, especially Christmas. We would exchange Christmas gifts and sit around bonfires, having a great time engaging in banter.

Sanuo and Ravannuo were our neighbours in Jalukie. Unlike Neibu, they were a mischievous duo and often kept Neibu engaged in their daily adventures. There were times when they would rush into our home and wake Neibu from a deep sleep. When he would wake up shocked, or angry, they'd have the audacity to say, 'We just came to wish you good morning!' and run away. That was the relationship they had as teenagers, and Neibu would give an arm and a leg for their well-being and protection.

This one time, we were all crossing a small bridge in Jalukie. Sanuo jumped over the bridge, and so did I and Akhrenuo. But Ravannuo, who was wearing a tight denim skirt, heard a ripping sound and realized that her skirt had torn. She screamed at Neibu as if it were his fault. He was walking right behind her, and it was dark. In response, Neibu asked her politely, 'How may I help you?' She said, 'Please walk in front of me.' And Neibu being Neibu politely and sincerely did as he was told, making sure she wouldn't feel awkward about it for even a moment.

For Sanuo, Neibu was her partner in everything. She would often joke, 'You're our spare tyre,' as whenever they were invited anywhere, she would take Neibu with them as her plus one. And he happily obliged.

As life went on, both my brothers had found their paths while I was still struggling to find myself and, more importantly, the person who was behind my origin. After all, I was only six months old when my whole life changed.

One day, I gathered up the courage and walked up to our father and asked, 'Where is Apfü (meaning my birth mother)?' Unfortunately, the answer I received was not the one I was looking for. Apuo said that our mother had never come to ask about her kids since the day she left home; therefore, it was only fair if we too stopped caring about her. With this statement, Apuo had clearly expressed his disapproval of us wanting to seek a relationship with our birth mother. He was, perhaps, carrying the guilt of dividing the family into two, and so wanted all his kids to have only one set of parents in the hope that they never felt that they belonged to a broken family.

However, despite what my father expected of me, my quest to see our real Apfü began. I had decided in my head and heart, without telling my brothers, that I would pounce at every opportunity I got to see my real mother in this lifetime. What I didn't know then was that—so would they.

4

Three Minus One

When Rüükielie left, Neibu and I were still in school but thought of ourselves as adults. After all, we had many younger siblings by this time, and we felt responsible for them. I was the proverbial good cop to all my younger brothers and sisters, who gladly left the policing to Neibu. I would chat and joke with them, while Neibu was strict. In many ways, he was the third guardian of the house, always keeping an eye on their studies and trying to make sure that no one indulged in any wasteful or harmful activity or misbehaviour.

Rüükielie, who had always been an extrovert, wouldn't come back home often. We assumed he must be busy with his friends and social circle in Kohima. Since Neibu and I anyway spent more time with each other than with him, it didn't affect us much. Although whenever Rüükielie did come home, we would enjoy his company and have a good time together.

One of the things that all three of us enjoyed doing together as youngsters was hunting. So, when Rüükielie would come over for the winter vacations, that would be one activity we would engage in for sure. Many times, only one or two of us would succeed in getting a catch. There is one time that I fondly look back at till date. I was in class nine, and Rüükielie had come over for his winter vacation. The three of us decided to go hunting early that morning. While the three of us were getting ready to leave, our stepmother teased us, saying, 'Today I will keep the water boiling and cook nothing else. I will only cook what you boys bring.' Fortunately, that day Rüükielie shot a pigeon, Neibu a squirrel and I, too, managed to shoot a quail. We came back home with our hands full and enjoyed a hearty meal. I can sometimes still feel the warmth of that happy night—our stomachs full of delicious food and our hearts full of the kind of love that only the togetherness of family brings.

At the time, the lifestyle of the youth of our state was undergoing a shift. They were beginning to get exposure to different cultures, which included their vices and the idea that one must be high to enjoy life! A small group of young people initially started using cheap pharmaceutical drugs, including some off-the-counter pharmaceutical syrups, to enjoy the feeling of a high. However, some locals believe, as travel and market interactions grew, stronger drugs, like heroin, started becoming available too. With time, the accessibility to such highly addictive substances only increased.

Since we were in the village, none of us, including our parents, were aware of what was happening in towns like Kohima. Our father, being the pastor, had brought us up with strong Christian values, and we grew up with the understanding that any kind of intoxication was forbidden. Hence, the idea that anyone in our family could use addictive drugs was unfathomable to us. Frankly, we didn't even know what drugs looked like or how they were consumed. Besides, since it was also just the beginning of the drug menace in the state, awareness of it hadn't penetrated villages like ours. So, until Rüükielie went to live in the town, none of us were even remotely aware that we might have needed to be worried about such a possibility.

Some excerpts from an article published much later, in the year 2020, sheds some light on what our state was dealing with at the time:

> The story of narcotics in Northeast India started around 1983 when heroin, a deadly derivative of morphine, started making an entry into our society, particularly Manipur. Within two decades, the Northeast states recorded 1,10,000 drug addicts.
>
> Proliferation of narcotics production in the world started in the Golden Triangle area following the Second World War. Almost 80 per cent of the world's production of opiates originates from the 'Golden Crescent' and the 'Golden Triangle' area, now called 'Golden Pentagon' with the induction of Vietnam—Cambodia and Nagaland—Manipur in Northeast India.
>
> Reports in 1989 pointed out that Manipur, Mizoram and Nagaland together accounted for the smuggling of at least 20 kg of heroin every day and that the bulk of it was sent to different parts of the country for eventual routing to the United States and Europe. Heroin was then sold under different brands such as 'Two Lions and a Globe', 'Double Globe', 'Five Stars', and even 'Dangerous'.[1]

When our brother came home for the holidays a few months after he had moved out, I noticed a peculiar change in him. I would often find him with a white powder, which he would consume in different ways. The three of us would share a room when he would come, so I could observe his activities closely.

Sometimes, Rüükielie would seem to eat the white powder, which he would carry in a cigarette pack. Other times, he would spread it out on what looked like a small piece of silver paper, heat it from below and 'smell' it. He seemed to become more relaxed after doing that and often just went off to sleep. It used to make me very curious.

During this time, Neibu had moved to Kohima to join the science college as well but was living with our relatives. Though Neibu was in Kohima too, his life revolved around college and home, leaving him with little to no time for interactions with Rüükielie. This was a fact we were not aware of, along with everything else. Which is why when Rüükielie told us that the white powder he was taking was medication for his pimples, we believed him. Regardless of the fact that we couldn't really see much acne on his face.

As time went by and Rüükielie got more and more trapped in the jaws of his addiction, it became hard for him to hide it, especially when he had to spend prolonged periods of time with us. Word reached our parents' ears through the people of our community, some of whom would visit the town for work and had got in touch with Rüükielie. Eventually, they intervened, and he was sent to a youth gospel camp organized by the church.[2]

Rüükielie, who himself wanted to quit the habit by the time this opportunity presented itself, readily agreed to go for it. Different speakers would come to the camp and teach the youth about religion, good habits and how to maintain the path shown by God. During his time there, Rüükielie did consciously work on himself and went back.

By the time Rüükielie's college years were ending, instead of continuing to study, he chose to take up a teaching job, in order to help our father with the running of the household. With this, he moved to Dimapur where he taught in a school. It was during this time in Dimapur, we discovered later, that there was also something else that Rüükielie did, the knowledge of which he withheld from us.

All three of us had never spoken of our birth mother. Being the youngest I didn't have any memory of her, and if my brothers did, I didn't know of it, because they never discussed it with me. It led to a lot of curiosity about her within me. We also understood by now, even though we were never explicitly told, that our father did not want any mention of her in front of our stepmother. As a result, she remained a mystery to me. But since Rüükielie had been the first among us to get out of the house, he had taken the liberty to go and meet with her. He had traced her; by this time she had remarried and had more children. He went to meet her without confiding in any of us.

Then, around the same time, again catching us unaware, his withdrawal symptoms got the better of him. Having completely stopped taking drugs, he would experience intense pain in the stomach, among other things. Since awareness of drugs and their after-effects was so limited, Rüükielie, who was convinced that his addiction was taken care of, was himself blind-sided; he thus wasn't prepared to handle the situation. When his condition got worse, some people from Jalukie, who lived near him and visited him frequently, found out and helped him get admitted to a hospital.

Nimbu Saab

Our birth mother, Nupulhuu, lived in Dimapur at the time, and since she was the only blood relative nearby, the people who admitted him informed her of his condition. When she received the news, she was making her way back home from church. Shocked to learn of her son's ill health, she quickly made her way to the hospital. As she laid eyes on her firstborn, suffering and in pain on the hospital bed, she felt helpless. In the rush to reach the hospital she hadn't informed anyone at home and had reached alone. Thus, when the doctor asked for some more medication, she had to leave her son alone to go and get them, despite how much she wanted to avoid being away from him. Rüükielie kept holding her hand and repeating, 'I want to go to Jalukie.' His constant pleas broke her heart. Ironically, despite being closest to the person who should've been like a home to him, he was feeling homesick. It thus chilled her to feel his desperation to go back to the place and the family he called home. It seemed as if he was scared and all he wanted to do was to rewind time. Holding his hand warmly and promising to be back soon, she almost ran out of the hospital, clutching the prescription in her hand.

When she hurried back to the hospital, she wasn't prepared for what she was about to see. A small crowd had gathered outside Rüükielie's room. However, there seemed to be no sense of urgency in anyone's movements. Instead, everyone seemed to be downcast. As she penetrated the crowd to reach her son's bed, her heart sank. Rüükielie, who until a few minutes back was wincing in pain, begging her to take him home, was lying lifeless on it now. In the little time that she had gone to get medicines for him, her firstborn had breathed his last. She shut her eyes tight, as if to shut herself off from reality. The tighter she squeezed her eyes, the more tears dropped down her cheeks. Soon they started flowing freely as the pain of losing her firstborn clutched at her heart.

The responsibility of collecting and delivering the mortal remains of her oldest son was now left to our birth mother. Once again, she went, along with her mother (our biological grandmother), who accompanied her to the house that had shut its doors on her firmly many years ago. How unfortunate was she that this time, she was taking home the dead body of the boy she had gone to ask for years ago. Somehow, they, along with a few other people of the community, reached our parents' house in Jalukie late at night.

My father and my stepmother were sleeping when they heard loud, restless knocks on their door. Not used to anyone visiting them at this hour, my father rushed to open it. Looking at his first wife standing on the other side of the door shocked him. However, she wasn't alone. Seeing that there were a few more people standing behind her, our father realized that something was wrong—but he could not have guessed how wrong.

The next few minutes were a blur, according to my father. From the minute he laid his eyes on Rüükielie, he could not stop looking at his son's face in disbelief. He kept at it all night. He was oblivious to everything else going on around him. After some discussions, the family and people present there decided that the last rites would only take place the next day, since it was too late at night when the body arrived.

Meanwhile, Neibu and I were in Nerhema. Neibu was visiting from Kohima, and I had my holidays too at the time. So, as was usual for us during those times, we had packed our bags and had gone to our grandparents. Now that we were older, we did not just play all day while we stayed there. We would help our grandfather with his paddy fields. In fact, every morning, both Neibu and I would first go to the fields and work there until afternoon.

It was a usual, sunny day in the fields, and both Neibu and I were working on the crops when someone suddenly called out Neibu's name from the hilltop overlooking the fields. We couldn't hear him properly, but he kept gesturing for us to go home.

Neibu was calm but alert and started rushing home while I followed him. We had no other information, but the way we had been summoned didn't feel right. When we reached home, we saw our grandparents huddled together in a corner, sobbing. The minute they laid their eyes on us, they started crying loudly. Between a lot of gestures and muffled words spoken with heavy throats, we understood what was being said.

We were being told that we had lost Rüükielie, that somehow his dead body had been brought to Jalukie by our birth mother and that we had to rush there for his last rites. I instantly turned towards Neibu, as if from muscle memory, for some kind of guidance. I couldn't understand how to react. My emotions sat troubled and heavy inside of me, but I didn't want to cry in front of my grandparents. I saw Neibu hold himself and take care of them and realized that I needed to do the same. We needed to be strong for them.

Neibu took the lead, arranging for a vehicle and preparing for the trip ahead. It was decided that only Neibu and I would go to Jalukie with the men, including the cousin who had travelled to get the news to us. Time and resources played their cruel part; our grandparents had to stay behind as the vehicle could only accommodate us. I will never forget the look on their faces, as the car pulled us away from them and they were left standing there, weeping but unable to do anything more than to just look at us leaving.

It was possibly the first time that Neibu and I had nothing to say to each other but a lot to understand from our silence. The three of us were supposed to see the ups and downs of life together. Rüükielie was supposed to be the one to get us all out the way he had often done in the past. I would look at Neibu several times throughout the journey but looking at his stoic expression, I couldn't gather the courage to talk to him about Rüükielie. Learning of him leaving us while we were in Nerhema was harder. Each turn of the road seemed to nestle a cherished story in its nook. Everything around reminded me of the three of us together. It was like having flashes of the times we had spent there, of another life, as we knew it. This was the place where we had first started being a pack of three—when Neibu and I would just follow Rüükielie around all day. It was the place where we had learned that while other things may change around us we three would never stop being a team. How could we? In the family that had grown so much larger over the last few years, my two brothers were the only ones I shared both birth parents with. If this was God's plan, it was a cruel one, I thought.

A thought that kept nagging me was this: It seemed that Rüükielie had all but slipped away in front of our eyes. That he had left us due to something that was so avoidable gnawed at me.

I hadn't even known that this was possible and the thought of it numbed me.

I now understood why Neibu always seemed so strict and particular about not just me, but all our siblings and the paths we chose. He was wiser than all of us, I thought. It had been getting easier and easier to find addictive substances in our state now, so I made a silent promise to never get drawn towards any of them and live just like Neibu did.

Although I was gutted, there was one thought that I couldn't shake off. I found myself going back, over and over, to the news about our birth mother bringing Rüükielie home. I had grown up with absolutely no memory of her. I didn't even know what to picture when I thought of her. I longed to see her. I longed to know the woman who had birthed me, and this may not have been the best time to get to know her, but I was very happy to settle for a glimpse. I could finally put a face to the idea of a mother, I thought. Thus, a small part of me was experiencing some happy curiosity, while also feeling a bit guilty at the thought.

Neibu, on the other hand, suddenly seemed to have grown up, as if he had aged well beyond his years within a few minutes. I could tell that he was wracked with grief, but he also seemed to be preparing himself to handle everything, now that he was the oldest surviving son of our parents. If he felt anything about seeing our birth mother too, I couldn't tell because he wasn't sharing anything with me in that moment. His expression remained serious and alert for the rest of that journey, as if he were planning for the next steps ahead. He was preparing to hold space for everyone's emotions but his own.

On the other end, my birth mother and her mother, who had accompanied Rüükielie's mortal remains, had spent the whole night sitting in a corner as they did not want to be a bother to anyone. My mother had to deal with her sadness while also navigating the awkwardness and discomfort of being in the house of her ex-husband with his new wife, and it was especially difficult because nobody was mentally prepared for any of this—neither for Rüükielie's sudden demise, nor for Nupulhuu's presence. By dawn, it was clear that they were neither needed nor would be welcomed. In fact, they were unwanted. Hence, not wanting to miss the one chance to take the next scheduled bus out of Jalukie, for they would have had to wait an entire day outside if they missed it, they left as soon as they could. Her heart broke at not being able to hug her other sons and share their grief, but she didn't want to cause any trouble for anyone.

Oblivious to her leaving and her reasons for it, Neibu and I arrived in Jalukie long after she had left. I had barely been able to let the reality of my brother lying dead in front of me sink in when my eyes started darting from one end of the room to the other, combing through the crowd, trying to look for her familiar face or for someone who could introduce me to her. As if right on cue, I overheard someone at a distance talking and mentioning how Rüükielie was brought in during the night by his mother who left early in the morning.

As soon as I heard these words, I felt a slight pang of regret, which then quickly turned to anger. Anger at my mother for not staying back to see us. She had just lost her firstborn, and this was her only chance to see her other two sons, I thought. Did she not love us at all? Was there no such thing as a mother's love? How ruthless was she? Why did she come all the way if she didn't want to even see us, let alone share our sadness, hold our hands or hug us warmly? Within minutes, all my curiosity and love for the mother who had birthed me turned to hatred. I hated her!

While I couldn't help but let my emotions against my mother get the better of me, Neibu, as we all silently expected from him, took on the role of the eldest in the family smoothly. He wept but also collected his emotions enough and in time to help move things on. Neibu and our father decided to bury Rüükielie in our backyard, and he made sure that all the ceremonies were carried out with minimal hiccups.

I will never forget the day when all of us stood around the grave of the oldest son of the Kengurüse family. I couldn't help but notice the contrast between my father and Neibu as they stood looking down at the ground under which Rüükielie now rested. Our father seemed to have aged a hundred years under the weight of the knowledge that he had lost his young son, his shoulders stooping low, and eyes swollen and red due to the steady flow of tears; he had the demeanour of a man who seemed to have had life sucked out of him in one night. On the other hand, Neibu seemed to have grown several years older in a matter of hours. Standing tall, his expression stoic, his arms around different family members from time to time, Neibu reminded them of the pole that he was to them.

Life as Neibu had known it, as a young carefree adolescent, was suddenly over, and I think he could never go back to being that boy again. He was, suddenly, a man.

5
The Grief

How does a man act the day after losing a loved one? What is the right way to grieve? Should you now regret all the things you wish you'd said to the person before? Should I have asked Rüükielie why and what he 'smelt' and ate off the white-and-silver cigarette paper? Since Neibu had never seen our one and only other brother drowning in these habits, maybe it had been my responsibility to question the 'medication' Rüükielie supposedly used for his pimples. Was it our dear Rüükielie's coping mechanism to deal with our parents' untimely separation? After all, he was the firstborn, and probably the most emotionally affected out of the three of us after our father remarried. I guess we will never know.

This was the first time I felt this extent of restlessness, with a hint of helplessness, surge through my mind and body. This was an emotion alien to me until now.

The morning after 22 June 1991, the sun rose as usual over Jalukie, but the Kengurüse household woke up to a reality none of us could've dreamed we would ever have to confront someday. A reality altered by grief.

We never expect a close relative to die before time—not until it hits home. No one in this world does, and we were no different. Until it happens to you, heart-wrenching incidents and accidents, such as this one, only happen in other people's families and homes.

There are no words to describe the morning after for the family, but if I had to truly describe that terrible dawn, it would be like this—the loss of Rüükielie hung heavily over us, like a shroud of sorrow that seemed to mute the vibrant greens of the paddy fields and the laughter of children playing in the village. Out of three potential breadwinners for our big family, now there were only two left. The weight of the responsibility felt heavier to me than ever before. Until then, it was just Neibu and Rüükielie who thought about providing our younger siblings with the basic essentials of life.

What will we do? How will we survive? Who will feed this large family of ours? Is this really, God's plan? How can it be? Who will teach Rüükielie's students in Dimapur? The fourteen-year-old me could not begin to answer all these existential questions.

Eventually, though, the 'moving on' began, as expected by society.

This may sound strange but considering our roots in Naga culture and history, we appreciate and worship all spirits—spirits residing in stones, rivers and trees—and every strand of nature's magnificence. Our ancestors would say that the power that makes one capable of acknowledging the world of spirits and the world of senses is 'true power'. The new generation of Nagas are disowning these ancestral beliefs and are often called foolish for doing so, for our folklore, vibrant as it is, is not merely a collection of stories; it is the root of our independence and identity, a treasure that, once lost, is irreplaceable. Since we're a joint family, with siblings ranging across age groups, we've been able to preserve this culture.

The Grief

Something we also believe in Naga culture is that the spirits of our ancestors stay with us forever, or long after they are gone, to watch over us in our day-to-day activities.

So, in keeping with these beliefs, after Rüükielie was buried in our backyard, we went back to the house and began to try to get our lives back on track.

Our house in Jalukie, made of bamboo and thatch, was just one long room divided into three parts, with a kitchen and one bathroom outside. We were already seven siblings (Kikrusenuo, Khrisaneisa, Keneitsilie, Neiketuü, Rüükielie, Neibu and me) by then, and most of us slept in one bed since we were too poor to have independent rooms. And then, there were three on the same bed—two once Neibu left for Kohima—and the youngest two siblings slept in Apuo and Apfü's room. Now, with Rüükielie gone, as the space on bed that I could use got bigger, so did the feeling of loss in my heart.

For days, after Rüükielie left us in the flesh, I believed he stayed with us in spirit. Every day, during that dark and gloomy phase for our family, I would hear in the silence a door screeching open and then shut. Someone, in the middle of the night, would be doing things in the kitchen. I would hear the cutlery and the utensils tinkling against each other—as if my departed brother had left something behind that he was now looking for, or was attending to some unfinished business. While praying to God that his good soul gets what he's looking for, I didn't utter a word to anyone. For the longest time, I believed only I could hear these sounds and that it was a secret between me and Rüükielie. So, every night he visited us, I'd say in my head, 'Oh my brother is back because he's missing us.'

However, there was one more person who heard what I did. Apuo. Every night, I'd find him near the grave weeping like a child, mourning his firstborn's passing. I would run out, console him, bring him back to his bed and go back to mine after.

These incidents, in which we believed Rüükielie came back to the house again and again, made me realize that we human beings have a soul that remembers even after death. This belief system helped me sleep at night sometimes.

The first death in my life and our family had a deep impact on all of us. But the men, the three blood-related men in the family after Rüükielie, being men, instead of grieving together, came up with their own coping mechanisms.

For instance, our strict father had a moment of profound introspection, which reshaped his approach to life and parenting. Days after Rüükielie's funeral, our father gathered us (Neibu, my stepmother and me) around the fireplace in the kitchen and announced a decision that took us by surprise.

'I will never scold you again,' he said, his voice strained with a resolve born from the deepest loss a parent can endure.

This seemingly simple statement was monumental for a man known for his strictness, a pastor who held discipline in the highest regard.

While I was left speechless after hearing his words, Neibu, ever the wise and thoughtful one, gently approached our father and said, 'If you stop guiding us because of this tragedy, how will we ever learn

The Grief

to stand strong? This is all part of God's plan.' His words, though respectful, carried the weight of a son who had suddenly found himself shouldering the responsibilities of being the eldest son. Once again, we, as a family, found comfort in looking towards the supreme power—dear God—to pave the way for us.

Neibu, on the other hand, was trying his best to turn this loss into a lesson all our younger siblings could learn from. Therefore, while he'd teach math and science to the kids at home, he would occasionally throw pearls of wisdom at them. 'Life is so short; we never know what will come anytime. So, you have to live well.'

'*Ache se rehena hai* (Live well),' he'd say. Every action of Neibu's was directed towards this life goal ever since he became cognizant of the state of our poverty-stricken family with its too many mouths to feed.

I do sometimes wonder if Rüükielie was the reason why Neibu stayed far from all toxic or addictive substances all his adult life. He had no vices. He would advise the siblings to do the same, but not all of us had the same willpower as him. His discipline in his personal as well as his professional life was unmatched by anything anybody in our family had witnessed before. That's why, his response to our grief was also very different from any of ours. He had the power to walk away from any situation in life with a lesson for himself.

While I was still grieving and looking forward to dreams that had Rüükielie in them, Neibu had developed the will to prevent any other such astounding incident from recurring in the future, for the sake of the whole family.

Nimbu Saab

We were all aware, especially Neibu and I, that the path ahead was not going to be easy; it never is when you lose someone so close to your heart. All we could do was try, and try we did. The foundations were laid strong in the fertile grounds of loss and love, faith and fortitude—eventually, the Kengurüse family, united in sorrow, picked up the broken pieces of their hearts and went back to focusing on surviving as one big Naga family.

Neibu returned to study science in Kohima. I went back to school for my tenth grade.

While I lost faith in God briefly, Apuo geared up to go back to serving in church as a pastor. Apfü and the rest of our siblings also went on with their respective routines.

After Neibu left for Kohima, I felt lonely and even powerless at times that I had to spend a whole year back in Jalukie, dealing with Rüükielie's memories, within the house and outside, before I could join him in Kohima for my higher studies.

Neibu's busy life in Kohima, perhaps, helped him cope better with our brother's untimely death. He also found a loving family with his local guardian, who was my father's first cousin, Aunt Medolhouü, a working woman with a government job, and her family, which consisted of her husband who was a contractor, and her four children—a daughter called Khris and three sons, Neithou, Hebuo and Ajavi. Neibu was the closest to Neithou and Hebuo as they were closer to his age. Neithou was the same age as Neibu and went to the same college as him, but as opposed

The Grief

to Neibu, who was pursuing the science stream, Neithou picked arts after his pre-university course. Hebuo, on the other hand, was three years younger than Neibu, closer to my age and as notorious as me. When I finally met our cousin brothers, I could see why Neibu cared for Neithou and Hebuo so much; they both were very much like Neibu and me in every way.

The first time the brothers Neithou and Hebuo saw Neibu sitting in their kitchen in his well-ironed formal pants and shirt, they had a thought, 'What a fine and sincere gentleman.' Their happiness spiked when they were told that Neibu was their cousin and was a student of science in the same college so he would stay with them for some time. Even though they had met this cousin of theirs for the first time, a few years too late perhaps, they looked forward to spending time with him.

While Neibu spent all the long weekends in Jalukie, he spent the short weekends with Aunt Medolhouü's family. Here too, just like back home, he behaved like a dutiful family member. He would wake up early and fetch water from the wells, which were 200–300 metres away from where they lived. He would even help the labour they hired while the house was being constructed, especially with watering the cement.

When he'd be back from college, while catching up with Aunt Medolhouü, he would occasionally say to her, 'I will become an O.' Then when she'd ask, 'What is O?' he'd smile shyly and say that the letter stood for 'officer'. Therefore, when he started teaching math in school later, she was surprised but glad that the job would give him experience in the real world and prepare him for any job.

While, as kids, we brothers felt the absence of a mother figure in our lives, Neibu had found three Apfüs in his—he had been lucky to have found that comfort in Aunt Medolhouü, whom he treated like a mother too. He would share with her all the problems from back home and even from college. She would advise him in every situation, and he would listen to her. It was her advice to Neibu, I discovered later, that led him to stay in touch with our birth mother. She suggested to him that it was not right that he didn't know his birth mother, but at the same time, she didn't push him to take any decision in haste. She knew that not forcing her opinions on him was key to their friendly relationship; so, instead, she'd ask him questions that would make him think hard.

With this aunt too, Neibu played the role of a humorous storyteller as he had done before at home for our parents; he would share anecdotes with her that made her laugh until her stomach hurt. Once, he narrated to her a dinner-table conversation from our house. He said, 'Whenever chicken would be cooked, the kids would be asked to choose which part of the chicken they would like to eat. Wings, thighs, legs, etc. Once, when some other meat, perhaps pork, had been cooked, the younger kids kept asking for wings. In response, our father grabbed a random piece of meat, plonked it on the complaining kid's plate and said, 'Here, this is your wing!'

Neibu's family life in Kohima was thus a mirror image of his life in Jalukie, sans the many baby siblings whom he had to tutor every day back home.

For Neibu, Rüükielie also stayed alive through his friends. During the time when he was in college, he had made new friends,

The Grief

and one of them was Rokomhalie (Roko), someone who was our neighbour back in Jalukie. He was Rüükielie's friend and went to the same church as ours. As Neibu was a year younger than him, Rokomhalie, like an elder brother, taught him math while in college. But their relationship was a great mix of mentorship and friendship. On days that were cold in Kohima, Neibu would rush to Roko's room and start singing '*Thandaa hai, chai garammm*' (The weather is freezing, give me a hot cup of tea) in the tune of '*Vande Mataram*', and follow it up with the line, '*Doodh bhi nahi … cheeni bhi nahi … khaile pheeka chai khai leee …*' (No milk, no sugar … let's just have bland tea) while waiting for his hot cup of tea.

Jokes apart—although they were great to keep us distracted—a question remained in our heads and hearts. We had always wondered how Rüükielie had ended up developing a drug addiction. To solve the puzzle, a few months later, we asked Rokomhalie if he knew the roots of Rüükielie's drug addiction. He seemed as lost as us. He said, 'I knew he was naughty, but I had no idea about this.' We were never able to solve this mystery, which gave us many sleepless nights around that time, but we soon learnt to live with it.

Another good friend Neibu had made in the college was Thepfuvilie. How they met for the first time involved a funny incident only Neibu could narrate well, but here, I'll try. One fine rainy night, in 1990, Neibu and Thepfuvilie were on their way to college from Kohima. They were both on foot as they both could not afford bus tickets. At that time in Nagaland, bus service was anyway the only mode of transport that most people could afford, and it was in sparse supply and not very cheap, hence youth from backgrounds like ours were forced to take lifts from truck drivers. More often than not, we would have to cover large distances, such as from Jalukie to Kohima, on foot. Luxuries like travelling in an Ambassador car, a Jeep or Shaktiman truck were beyond our means as well as imagination. So, one day, while walking to college, which was 10 kilometres away, both Neibu and Thepfuvilie happened to take shelter at the same spot, below a rock. The conversation began in Nagamese, the language that everyone around us generally uses. But on discovering that they both belonged to the Angami tribe, they had, almost as if in reflex, switched to the Angami language. And ever since, they were inseparable. The cherry on the cake was that their rooms were also adjacent to each other. They studied together and went to church together; in fact, Neibu convinced Thepfuvilie to go for his first church service.

It appeared to me that college made Neibu feel more confident and stronger in his character. His family background no longer made him feel weak; he was finally feeling like an equal with the rest of his class, no matter where they came from. Thepfuvilie shared stories of Neibu's 'short temper' with me, a character trait none of us, his family and friends back in Jalukie, were familiar with. For instance, he narrated to me one incident from the time when Neibu was already the president of the Tenyimy Students' Union. The role was even bigger than that of the president of the Angami Union, so he had authority and was proud of it. During this time, on one occasion, Neibu, Thepfuvilie and another classmate of theirs were discussing their freshers' programme, keeping in mind the fact that after class, typically, students tended to be stressed and edgy, and just ready to explode. Mid-conversation, there was an

The Grief

argument with the classmate and Neibu said to him, 'Since I am the president, please follow what I say.' Neibu's body language and assertive tone suggested that he could beat up the classmate any minute, so Thepfuvilie intervened and mediated the conversation instantly.

As Neibu was transitioning from a teenager to a strong, independent, audacious man, and even though he was now a science student, his love for spirituality remained and probably grew.

I was happy for Neibu; his life in Kohima seemed exciting and it was becoming more and more tempting for me to follow in his footsteps. But that life for me was still a year away. So, while Neibu had new people, things and experiences to look forward to, my life in Jalukie had come to a standstill. I was temporarily stuck in the same place I had spent most of my life in, with my brothers—only this time, I was without my partners in crime.

Back in Jalukie, as much as I tried to drown my miseries in my school and studies, my marks didn't reflect the hard work I had put in, against all odds. Unfortunately, the loss of my big brother and the sudden weight of responsibilities at home didn't bode well for my academics. It broke my heart and confidence. I should've listened to my brother Neibu when he pushed me to pick up my books and study!

I vividly remember picking up my marksheet at the school in Jalukie after my matriculation and without even reading it, taking a bus to some place in Kohima. I had two data points that I went confidently with to the big city—one, Neibu studies in the science college and two, he resides in Lakeview hostel.

With this information, I reached the right place and met Neibu to show him my marks. I had scored only 40 per cent in the two subjects that were the most important to join the science stream in college—math and science. To my surprise, instead of scolding me and giving me a lecture, Neibu congratulated me for passing my exams, considering our situation the previous year. However, we both knew the harsh truth. With such marks in math and science, I would not be able to study science in college. Neibu even sighed and said, 'Had you got even 45 per cent, I'd have spoken with my lecturers and requested them to give you a seat.' Having said that, he suggested I now pick commerce instead as my main subject. I, being the stubborn one, said no. I confessed to him that if I could not get into the science stream in his college, I'd rather take up arts in some other college than break my head over math again with a subject I didn't want or understand—commerce. He listened to me and then guided me towards colleges that offered good courses in arts.

Finally, after a whole day of up and down, round and round, I got admitted to Alder College, which was located around 20 kilometres away from Neibu's college. I felt safe again, with my elder brother in the vicinity. Even though we didn't live in the same house anymore, and my college was on the other side of the town, I felt at home again, after a long time.

I loved being back in the shadow of my talented and extraordinary big brother. He was again ahead of his class in everything, like a true

The Grief

leader. Akhrenuo, Neibu's childhood crush, who was also studying as his junior pursuing BSc in nursing, would often tell us that everyone knew Neibu. In the science college, he was in the Evangelical Union, the president of the Angami fellowship and, as you may remember, eventually, the president of the Tenyimy Students' Union. He was also responsible for many of his friends in science, including Roko and Thepfuvilie, doing their first church service.

With age and accomplishments, Neibu was also becoming a charmer, and so, every girl in his class looked up to him. Gone were the days when he was looked at by more privileged and wealthier classmates and laughed at for his 'poor-man' dressing sense. If there was one person that I knew who could set his mind to something and make it happen, it was Neibu; he single-handedly made sure he changed the way people looked at us as we grew older.

Considering his interests in science and spirituality, everyone around him believed that he would opt to become a doctor after graduation, or if not that, he'd be an evangelist or perhaps, a political leader. Even though nobody could put a finger on what Neibu truly wanted to pursue for a career, they all knew that whatever he did would change the lives of his family and friends massively. That was the impact he had on their lives.

Seeing Neibu shine bright everywhere he went, I was slowly beginning to get inspired, too.

I had already suffered by not studying in Jalukie despite being pushed by my wiser older brother; I would not dare to repeat that mistake.

As life got busier for me and Neibu—with our respective struggles of making ends meet, studies, co-curricular activities and

Nimbu Saab

friends—and as we made a place for ourselves in the big city of Kohima, many times we found comfort in the fact that it was just us, as a team, against the world. This sense of moral support made us feel less alone and more motivated to follow our dreams.

6

My Brother Neibu and Me

As soon as Neibu finished his graduation, he got a call from our father. Apuo had been sponsoring us with the modest salary he earned, so right after Neibu finished college, he decided that Neibu now needed to share the load. Our father asked him to come home and help him out.

As Neibu was packing his bags and winding things up to return to Jalukie, he was approached by people from a reputed school called Baptist High Senior Secondary School in Kohima, inviting him to come and work for them as a teacher. Neibu, who enjoyed academics and was used to teaching our younger siblings as well, didn't need time to think. Other reasons aside, he knew he could use the money. Thus, he readily agreed.

Up until now, our father had been paying Neibu's hostel fees as well as my room rent. I was done with my pre-university course and was pursuing a college degree in Kohima at the time. Since Neibu was staying in a hostel, his fee covered his food, but for me, after taking the rent out, I was left with very little to afford three square meals a day.

Neibu's job saved me too.

Sometime before Neibu was called, I received a call as well. My father told me that his salary was not enough to cover both our college expenses so I must leave. I resisted, saying I deserved to at least study till the same level as my brothers had and held my ground. In response, my father simply said that in that case, he could only afford to send us both a sum of ₹650 each. As Neibu's fee of ₹450 covered all his basics—including accommodation and meals—he was left with ₹200 in hand to buy other essentials and utilities. On the other hand, I was living in a rented space, and so after paying the rent and college fee of ₹300, I would only be left with ₹350 for meals, utilities, travelling and other miscellaneous expenditure. Rice alone cost ₹150 for 15 kgs, which was the minimum that we needed in a month. Our budget was so tight that if a friend ended up accompanying me for dinner, I had to make sure that my next meal was at his expense, or else I would run out of ration. At the end of the month, I used to be left with barely ₹200 to cover everything else. Hence, on Saturdays, I even started going to Nerhema to collect firewood and to sell it in Kohima with the aim of making some extra money. In fact, in those days, when Neibu would sometimes visit me, he would come with some supplies, like four to five kilos of potatoes or five kilos of rice—whatever he could spare, given that he too was on a tight budget—and leave them for me. After surviving like that for a year, I made up my mind to run for the college elections just to be able to make some money, as cash came with the appointment. Luck favoured me when I won them and was appointed the general secretary of my college.

However, with his new job, Neibu needed to move out and rent a place for himself—somewhere he could work from. So, as soon as he had found a suitable place and had moved in, he came knocking at my door, demanding that I come live with him, so that our father's load could be lightened. Happy to be reunited with my brother again, I promptly agreed and followed him.

Neibu started teaching after this, and almost as soon as he began, he started making a mark for himself everywhere. He had the unique ability to teach every child, based on his own ability, and that earned him a lot of admiration and respect. Aside from teaching, he also had a commanding presence and his students would fall in line on their own, out of respect for him.

Soon, he started getting approached by many parents who wanted him to take private tuitions for their children. Neibu, who was desperate to earn more money, thought of it as a great way to increase his income substantially and started taking private tuitions in the apartment we were sharing. He converted one room into a makeshift classroom, with chairs and a board, and would start teaching from as early as 5 a.m., followed by classes at school and ending with tuitions again at home.

Meanwhile, I wasn't as interested in studies as he was. I knew I had to complete them, since we all understood that that was our only channel to a better future. But I wasn't as serious as Neibu. I liked enjoying my life. My only responsibility was to cook for us. So, after I did this, off I would go to hang out with my friends and take part in my college activities. I would often look at Neibu and wonder how he could keep going day after day, toiling all the time and not complaining about it the least bit. In fact, each chance that he got, he would sit with me and explain how he thought that I was capable of much more and could do so much better if I worked harder.

A few months after we had moved, our aunt, who was worried about our cousin Hebuo's studies, called Neibu. He had to take his matriculation exams that year, and our aunt was worried that he might not do well. Neibu offered lessons for students of matriculation too, so she asked him to keep Hebuo with him and teach him too. In fact, our uncle, the stricter of the parents, made it very clear that Hebuo's results were the responsibility of Neibu as much as it was Hebuo's. So, Hebuo packed his bags, and we got our third flatmate.

Those days, Neibu was busier than ever with little to no time left to do anything else. His work did pay well by this time, but since most of the money he would get was often used up in meeting our own and our family's needs, reducing his workload or work hours was not an option. I would rarely see him relax with his friends, although I knew he enjoyed doing so. All through college he had a circle of friends who were with him, and some from Jalukie who were now in Kohima would also meet him as often as they could, but now with his routine, he couldn't afford these pleasures.

But I knew Neibu had found his own ray of sunshine in the midst of it all. After teaching in the Baptist school for about six months, Neibu had been appointed by the government and had joined a government high school in Kohima. Soon after joining that school, I could see a change in Neibu. He had met a beautiful girl, Caroline (name changed), who taught with him in the school. Neibu and I didn't share the kind of relationship where he would sit and tell me what was happening in his love life in detail, but I started noticing that there was something different when he started making references and talking about this one girl more than usual. I also happened to overhear some students tease the two of them while they were in school. Her name always left a smile on Neibu's face, and I could sense that he was in love.

Soon, Caroline started coming over every once in a while, and while they would cook and spend time together, I would, as usual, find an excuse to step out and let the two of them have time alone. Visits to each other's places, with me and our cousin in tow, and cooking together became common, too, though we couldn't do it too often as Neibu's routine didn't allow it. But my brother was happy, and I felt happy to see that.

On the other hand, while I had many friends including girls, I didn't really have a special someone. Though my heart did beat for one girl a little more than it ever had for anyone else. Her name was Kesovonuo, and she was our neighbour in Kohima, although that is not how we knew her. Kesovonuo or Aseü, as she was fondly addressed by most of us, grew up in Nerhema, where her family lived next to our grandfather's house. Every vacation, when my siblings and I went to Nerhema, one of the first things we used to do was to dump our bags and pick up empty pitchers to run to the common well for water. This is where both Aseü and I noticed each other for the first time. Many people of the village used to find a large family like ours, with so many of us siblings, rushing from here to there together rather amusing. But our family was also admired for being so tight-knit despite being so large. Aseü and her family also shared this view. Aseü was also a Kengurüse, which meant that we were related, albeit distantly.

Nimbu Saab

While we had been in the same vicinity since childhood, our paths only crossed when we began living in houses opposite each other in Kohima. She had moved there for her pre-university course and since we knew of each other from the village, she would sometimes ask for my notes. In fact, she took some lessons from Neibu, whom she admired a lot—just like everyone else did. One time she asked him how she should address him—whether she should use 'sir' or his first name. He promptly responded with an 'Oh, please never call me sir—ever', making her become even fonder of him. She also felt a certain sisterly affection towards him since back in the village her elder sister and Neibu would spend time practising Christmas songs together.

However, the relationship Aseü and I shared was far from the loving, cordial one that she and Neibu had. In fact, often we couldn't go more than ten minutes without arguing. We had a quintessential love-hate relationship. But for some reason I really trusted her and loved one thing about her—she really cared.

When I would be away, visiting the village, leaving Neibu and our cousin behind, it was Aseü who would come over and cook for them. Many times, she would come over just to make a cup of tea and always make sure that they were taken care of. In fact, at one time, when our sister Neiketuü was visiting, and I was away, Neibu called Aseü and told her that he wanted to buy Neiketuü something nice as he had just received his salary. Aseü graciously offered to accompany them to the market, and the three of them bought a beautiful, purple velvet dress, which Neiketuü later wore for Christmas.

Sometimes in the middle of our banter, I would just suddenly blurt out 'I love you', to which she would roll her eyes and say, 'And I hate you', and we would leave it at that. We weren't an official couple, but within our hearts we both knew that we weren't just friends either. Yet, neither of us seemed to feel the need to figure things out during that time in our lives; so aside from some sporadic teasing, we didn't discuss our future. Besides, Neibu seemed to be worried about me and would keep checking on me, and asking me to study, to not be out so much or do things that would take my focus away from studies. Sometimes, it would even irritate me, or make me argue.

One such thing that troubled Neibu were the late nights I would enjoy sometimes. Every time he saw me do it, it would always be followed by what I thought of as a lecture at the time. I know he always worried that I, or for that matter any of us siblings, would follow in Rüükielie's footsteps, as with every passing year, the access to substances only kept getting easier. But I too was a god-fearing Christian and never even thought of indulging in such things, so his concern would sometimes seem, to me, to be a lack of faith in me, irritating me.

One time, after one of our arguments, I was feeling disgruntled. In my anger, I told Neibu how ungrateful and thankless he was since I cooked and cleaned for all three of us and he still thought nothing about scolding me. I wanted to make him realize how it would be if I didn't do what I was doing for them routinely, so while cooking meat for their meal later, I had an idea and left it a bit uncooked. I shut the stove and walked away. Hebuo later told me that when they sat down to eat, as soon as they had sunk their teeth into a piece and realized it hadn't cooked through, they immediately understood that I had done it deliberately and had ended up breaking into peals of laughter. All my hopes of making Neibu feel sorry for his words came crashing down. Little did I realize that such actions had no effect on Neibu, except for drawing laughter from him. Just like him, his thoughts weren't frivolous and so he always thought everything out before saying it. He knew he was scolding me with the right intention and if I acted out, it could only make him laugh out loud; there was nothing for him to realize in the first place.

But I was young and my antics were the only way I knew how to deal with the anger that I would sometimes feel. Another time, when I came back home after a movie, much later than usual, Neibu and Hebuo, who were usually up at 4.30 a.m. every day, had gone off to sleep. I rang the bell a few times, but Neibu refused to open the door, and Hebuo was too scared to go against Neibu's wishes. So, the door remained locked from inside. I was furious. While I was looking around to grab something to open the door with, my eyes fell on the washed clothes that I had put out for drying earlier that evening. It was as if the mischief bulb inside my head had just been switched on. I went over to the railing, picked up all the freshly washed clothes and laid them all down on the dirty floor in front of the door of the house. I put some on to keep warm after the 'bed' was made and lay down right in front of the door. When Neibu opened the door in the morning, he simply rolled his eyes and asked me not to be so late again in the future, leaving me feeling silly for my act.

Sometimes I feel that since Neibu, around this time, had Caroline as a part of his life, he seemed happier and in better spirits. More understanding than impatient, and I liked that side of him. I thought she brought out the best in him. However, there was one issue between them. We were Baptists, while she was Catholic, and given how religious he was, that difference did matter to him. Be that as it may, he wanted to make it work, and that was clear to me because of what I found out a few years later.

Even though Neibu and I never spoke of our birth mother, even after Rüükielie's passing away, I know we both thought of her. I, too, never shared my disgruntlement towards her with him since Rüükielie and Neibu were the ones who had some remnants of memory of her, and I was too young to remember anything. With Rüükielie gone, Neibu couldn't find the space to discuss what he thought about her with me.

But I know we both wondered why Rüükielie never told us that he had been meeting her and thought about how ironic it was that the last time she had come to my father's doorstep was to ask for her sons and the next time she did, it was to return one of them, albeit lifeless. My dream of getting to know how the mother who had birthed me looked was crushed when I found her gone during Rüükielie's last rites, and my feelings were so strong that I felt a little uneasy sharing it with anyone.

Little did I know that I was in for a surprise. One day in Kohima, while Neibu was teaching his students in the room and I was cooking our meal, I heard what I thought was the voice of an old woman, speaking in Nagamese (unlike us who usually speak in Angami), calling out to ask if anyone was home. I saw an old woman standing with a young girl, who I assumed was her daughter, saying, 'Anyone at home? Is there anyone at home?' I quickly rinsed my hands and was still patting them dry on the sides of my pants when I hurried towards the door to open it. As I was pulling the door open, the anxious woman looked at me and said, 'Neibu?' I smiled knowingly, since this wasn't the first time that someone was mistaking me for him and invited her and her daughter inside, saying, 'Oh, you are looking for Neibu? Please be seated, I will just call him.'

On realizing that I was not him, the woman suddenly held my hand, as if to get a better look at my face and suddenly started weeping. Thoroughly confused, I felt my gaze shift from her to her daughter and back to her. As I too looked intently this time, I realized that I had been missing something obvious. Her face looked very familiar to the one I was used to looking at every morning in the mirror. More than anything else, there was a big, telling mole in the middle of the brow that both of us had in common. It struck me immediately, and a bit uncomfortably, that this woman was not a stranger. It was in fact the one who had birthed me, my mother.

My doubt was confirmed when I heard the words she was managing to utter between her sobs. She kept saying, 'My son. My little son couldn't recognize me,' while she cried some more. I kept looking at her too and thought to myself, 'Oh! This is my mother. Finally, this is my mother!'

Neibu, on hearing the noise, came outside to look. Looking at all of us together, he decided that his attention was needed, and so, since his class was already about to end, he dismissed his students, making sure they all left quickly, and sat down with us. My mother continued weeping and feeling sorry about us. She kept saying how unfortunate it was that her son, who was all of twenty years old, didn't even recognize her. For some reason, she kept apologizing to me.

On the other hand, I had always imagined that if, and when, I would meet my Apfü I would be angry. I had thought that I would have so many questions to ask, and that I would demand to know how it was so easy for her to not even see us all these years and then go back home without seeing us when she had come with Rüükielie's remains. There were so many scenarios that I had imagined and played out in my head over and over. But, strangely, now that she was sitting in front of me, I felt numb. I felt devoid of any anger or any other feeling. Surprisingly, I felt nothing.

Neibu went on to ask them both to join us for dinner. Which they did. Once the initial shock of meeting my birth mother suddenly wore off, I realized that there was something else I was missing too. Clearly, Neibu didn't share my feeling. There was no sense of shock in the way he was interacting with our mother. On the contrary, there seemed to be a sense of comfort and familiarity. It seemed that both were completely aware of the ongoings of each other's lives. The girl accompanying our mother was her daughter and our half-sister, and Neibu already seemed to know that. In fact, the purpose of our Apfü's visit was to approach Neibu for help with the admission for one of her sons into school. Which meant that she already knew that Neibu was a schoolteacher.

The rest of the evening went by rather uneventfully. Our mother stopped crying eventually and all of us settled into a comfortable conversation over warm food and a new sense of familiarity. However, during the course of the evening I realized, uneasily, how I just felt nothing.

Later, when I spoke to Neibu, I found out that not only had he gone to meet our mother, but he had taken Caroline along with him as well. He told me how on one of his visits, our aunt, Hebuo's mother, had spoken to him about our mother as she knew of her; our aunt and our mother both worked for the police—as a clerk and a class four worker, respectively—in the same department. She told him how she felt it was imperative for a child to know the woman who had birthed him, and that she too deserved to know him. Knowing that Rüükielie had already re-established contact with her gave Neibu some push as well. Thus, when he found out that our birth mother was undergoing a medical procedure and was admitted in a hospital in Dimapur, Neibu had taken Caroline and gone to meet her. He had told our mother that he had brought the girl he wanted to marry, to meet her. Our birth mother was overjoyed. Her husband and children too were supportive and so, Neibu and her had been in touch ever since.

I would be lying if I said that I never thought of why Neibu didn't tell me. Was he worried that I wouldn't approve, or did he simply want to let me make my own decision about it? I would never learn the answer to that since I never got around to asking him. Instead, encouraged by the step taken by him, and Rüükielie before him, I followed suit. In my heart of hearts, I was grateful to them for paving the way and giving me the silent nod of approval that I had been waiting for. I too started visiting my mother and getting to know my half siblings whenever I could. Often, when

I would look at my mother, I would be amazed at how similar the moles on our respective foreheads were and would also feel silly about completely missing it during the first few minutes of meeting her. I would think how, growing up, I had never even seen a picture of hers, and there was absolutely no mention of her in our house. It seemed like there was an unsaid rule that forbade any mention of our birth mother, especially in front of our stepmother. But once I met her, knew how she looked, clearly saw the strong resemblance that we both shared, I would often find myself being enveloped in the warm comfort of finally knowing who I belonged to. I felt thankful for the fact that her new family didn't seem to have any rules in their home like the ones we did in ours. I wished I could thank Rüükielie, for in some way, it was him who gave me the gift of my mother after so many years, but of course it was too late to do that. I would tell myself that God wanted it to be like this. That perhaps, this was Rüükielie's purpose, and he had fulfilled it in his life and even in death.

Life may not have been perfect, but it did seem to be getting better now. The recognition of Neibu's teaching abilities kept spreading steadily, and he would have parents come in from everywhere, requesting him to take their children under his wing. Offers weren't limited to tuition only. Every once in a while, he would also have principals of other schools approach him to come and teach for them, even if he could only manage to teach a few classes.

Even though Neibu's plate was more than full, he never stopped working on himself. The money was coming in steadily, and he was making a fair amount by himself, but for a family as large as ours, there still wasn't enough to spare. Neibu realized that, too. Besides, it wasn't just money or the comforts that it could bring that he was hungry for. His hunger extended to a lot more. He wanted respect. He wanted stature. His work as a teacher did get him respect, but it wasn't the kind that he craved. He wanted it to come with better status in the broader society too. Not just for him, for his family as well. He wanted to uplift us all. Thus, when he wasn't teaching or sleeping, he would be found studying.

As a result of his busy routine, his trips to Jalukie kept getting less frequent. He worried about that too, since he also wanted to ensure that the rest of our siblings were working hard and stayed on track as well. We did not have mobile phones, or even landlines, in every house back then, so the only way to get updates was through letters or visits. Keeping this in mind, he made sure to make visits or send us letters every chance he got.

Although the younger ones were all a little scared of him on account of how strict he was, they enjoyed seeing him, and also looked forward to him arriving with the treats that he would carry for them every time. He loved to find new tuck or biscuits in the market and would make it a point to carry them with him for his younger siblings to try them. Everyone at home would excitedly look forward to the little surprises that Neibu had to offer them.

One time when he came home, he was carrying a few packets of something that looked like chunky bits of chips or possibly biscuits. No one knew what it was since no one had seen anything like that before. Neibu was conscious of the younger ones' nutritional needs too, and so had picked up a few packets of something that claimed to be packed with protein. At home, once everyone got together, he made them all sit around and distributed the packets, to be shared equally by everyone. The youngest ones were always supposed to be the first ones to try whatever he got. He wanted to encourage them to develop their tastes, so he would also ensure that no one missed trying whatever was being offered to them.

This time again, as per his usual habit, he followed the same routine and waited to see everyone's reaction. Everyone tried taking a bite of the chunky food item and seemed to struggle with swallowing it. However, none of us complained since we also didn't want to upset Neibu. Confused by the funny expression that had now appeared on everyone's faces, Neibu decided to take a bite even as he continued encouraging us to try 'the different kind of biscuit'. It took him a few seconds to realize that the 'biscuit' was quite hard and mostly tasteless. That's when he looked at the packet properly to see what it was. To his horror, he discovered that he had carried packets of soybean chunks—something like Nutrela—assuming it was a nutty variation of a biscuit! They were supposed to be cooked before consumption and here he was, making his family eat them in their raw form. He almost choked on the bite and so quickly asked all the others to spit it out and give the packets back to him. Not used to seeing Neibu be wrong about anything, this incident became the go-to joke between us siblings every time any of us ate something unpalatable.

These trips would usually last for a day or two, after which Neibu would be on his way back. In his quest to find new career paths, Neibu had also started talking to different professionals from outside the community, especially those who were posted in Kohima or around. Two such advisors he was speaking to were Col

Robertson and Col J.K.G. Kutty. The former was a public relations officer. Neibu looked up to the two of them, and they too were very fond of Neibu. That was perhaps his first interaction with army officers, and he was fascinated with how much respect the uniform commanded.

This was the time when Neibu started thinking of taking up a profession where he could command a similar kind of respect. He realized how they were looked up to and how the stature extended to their family as well. Perhaps that is what he needed. In fact, he even encouraged his friend Ravannuo, whose house was right below Col Robertson's, to meet with him and see if she would also want to learn how to join the army. However, Ravannuo wanted to follow a different path and that story ended there.

Neibu was a realist. Which is why when he was preparing, he didn't put all his eggs into one basket; he appeared for more than one exam.

Caroline was with Neibu every step of the way. She supported and encouraged his dreams. I liked to see him with her. With us too, she was always very polite and respectful, and so my respect and admiration for her only grew. I was truly happy for Neibu. I felt he deserved to have a piece of rainbow in his otherwise grey life where he was constantly pushing his way through his circumstances—making way for not just himself, but also the rest of our family.

Finally, the day that he had been waiting for came and his results were announced. As we all already expected, he had made it. But not just in one exam. He had made it in no less than three! Neibu was elated, as were we. We may never have said it openly in front of each other but all of us siblings and our parents looked towards Neibu to give us a better future. We, in our heart of hearts, believed that one day Neibu would carry us all to better times. This day definitely felt like that.

The three exams that Neibu had made it through could give him a job as a bank provisional officer; in life insurance; or, lastly, in the Indian Army, through the Combined Defence Services (CDS) exam. When he started telling our family and people around him, he realized that the decision-making process wasn't going to be so easy. His feeling about the army wasn't shared by everyone in the community. Especially people of our father's generation. They were the ones who had lived through the times of struggle with the Indian Army and still harboured some anti-India feelings. Our father knew this very well and was conscious of how some of these people from our community would think. We are all very close-knit, and he didn't want Neibu to be thought of as different. Besides, a life in the army also meant putting your life at stake every day, or so he understood. So, when he was told of the results, he wasn't as happy about his son's bent towards the army as Neibu had hoped. In fact, most others too encouraged Neibu to take up one of the other options.

Aside from a handful of supporters, Neibu's own will to don the uniform and experience the pride of doing that won the day. His dreams were bigger than for just a stable, respectable job. He wanted more—for him and for his family. More than anything else, he wanted to push open the doors of a brighter future for the rest of his family. He wanted to give a dream to each of the younger ones too. He knew they all needed greater exposure to be able to open their eyes to a better and bigger life before they could dream of one for themselves. He knew he needed to lead by example, and an example is what he decided to set.

Determined as he was, there was one big roadblock that bothered him. The family was at a crucial juncture; when all of us, aside from him, were studying, and someone or the other was in each of the quarters of the usual educational journey. Neibu had been the primary sponsor for all of us and the reason for the relative freedom we had, to be able to choose what we wanted to do and not be limited by what we needed to do. And this included me.

Choosing to join the army would mean going away, for a year and a half, for training before he could start sending money home again. Due to the precarious financial position that the family was in, Neibu's career move thus translated to a big, almost non-negotiable, issue to deal with. Could he choose his career over the sacrifices of one or more of us? Used to playing the role of the provider, caretaker, mentor and guide for all of us, Neibu did not have the liberty to take a decision independent of our interests. He wouldn't allow himself to either.

I could see that Neibu was worrying himself sick about his decision. I could see how conflicted he was. Perhaps we hadn't even thought that clearing the exam was going to be the smaller challenge. The bigger one was to navigate through this time while keeping everyone's best interests at the centre. Neibu, being himself, didn't want to burden any of us with his thoughts either, so he wouldn't share them with me or any of us. But I knew my brother. I knew the thought of not being able to provide for us,

Kengurüse family picnic.

GC Sundeep Khatri and Neibu in a bunker at the forward post in Naushera, J&K, during on-the-job training in November 1998.

GC Sundeep Khatri, Neibu, GC Vishal Singh Bais, GC Sachin Duseja and GC Suresh Dhaka leaving for on-the-job training in Jammu, during their third term in IMA in November 1998.

Neibu and friends just a few days before the passing out parade.

From left to right: Neibu, GC Rajeev Bhargav and GC Vishal Singh Bais at the farewell thrown by second-termers at the Thimayya cadet mess, in December 1998.

During on-the-job training at a forward post in Jammu sector in November 1998. In the picture: Neibu, GC Vishal Singh Bais and GC Muthu Ganesh (Sangro Company, Third term, IMA, Dehradun).

Neibu and his buddies enthusiastically working out in the gym during games session at the beginning of their second term at IMA, Dehradun, in February 1998.

Neibu, GC Ajay Dabur (on Neibu's side of the table) and GC Vishal Singh Bais, during their second term, in March 1998, enjoying lassi in Dehradun town during their free time.

Neibu's father Neiselie, stepmother Dinuo and sisters Arhenuo and Mhozienuo walking towards the place where Neibu was buried.

Lt Rauthela (in uniform) with the traditionally dressed Naga youth leaders who escorted Neibu's coffin all the way from the Dimapur airport to Phezha.

People from different villages from across Nagaland waiting in line for Neibu's coffin (the queue extended up to 2 kilometres) to arrive at Phezha, Kohima.

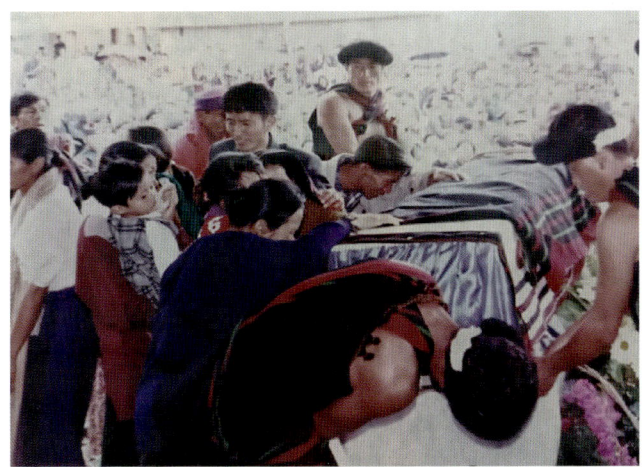

Chief Minister Neiphiu Rio at Neibu's wreath-laying ceremony.

The Kengurüse family (with brother Atoulie in the dark grey coat) receiving Neibu's coffin from the Indian Army.

Neibu's portrait adorning the wall of his brother Atoulie's petrol pump in Dimapur.

The hilltop close to their grandparents' house in Nerhema where Rüükielie, Neibu and Atoulie spent hours playing together as kids.

Just before Neibu's father (bottom right) prayed for both Neha and Diksha (centre bottom in black and white, respectively) at their home in Kohima with the rest of Kengurüse siblings.

Pages from Neibu's personal diary

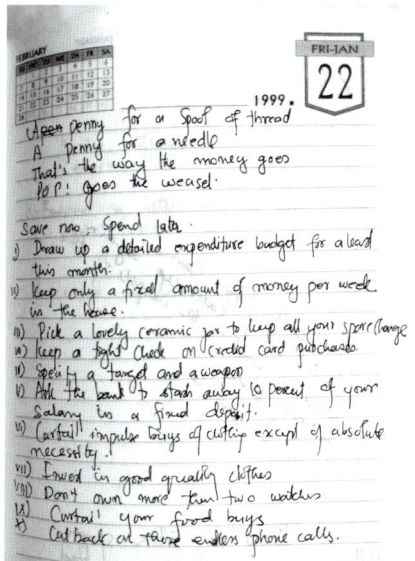

FRI-JAN 22, 1999

A penny for a spool of thread
A penny for a needle
That's the way the money goes
Pop! goes the weasel.

Save now, Spend later.
i) Draw up a detailed expenditure budget for at least this month.
ii) Keep only a fixed amount of money per week in the house.
iii) Pick a lovely ceramic jar to keep all your spare change
iv) Keep a tight check on credit card purchases.
v) Specify a target and a weapon
vi) Ask the bank to stash away 10 percent of your salary in a fixed deposit.
vii) Curtail impulse buys of clothing except of absolute necessity.
viii) Invest in good quality clothes
ix) Don't own more than two watches
x) Curtail your food buys
xi) Cut back on those endless phone calls.

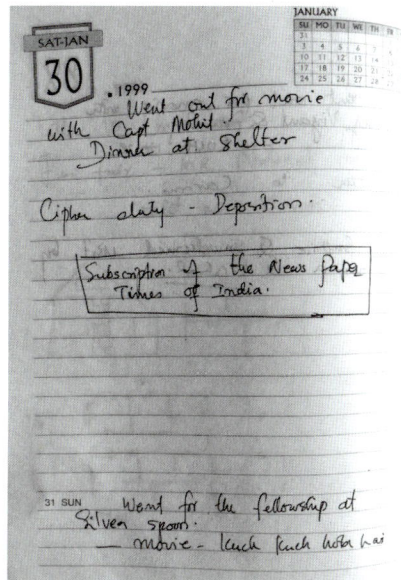

SAT-JAN 30, 1999

Went out for movie with Capt Mohit
Dinner at shelter

Cipher duty - Deposition.

Subscription of the News Paper Times of India.

31 SUN
Went for the fellowship at Silver spoon.
— movie - Kuch Kuch Hota Hai

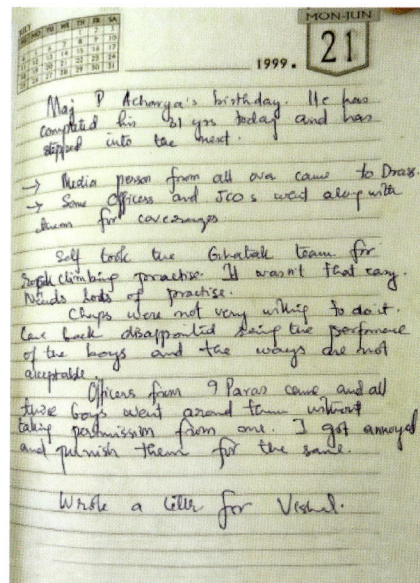

MON-JUN 21, 1999

Maj P Acharya's birthday. He has completed his 31 yrs today and has stepped into the next.

→ Media person from all over came to Draas.
→ Some Officers and JCOs went along with them for coverages.

Self took the Ghatak team for Rock climbing practise. It wasn't that easy. Needs lots of practise.

Chaps were not very willing to do it. Came back disappointed seeing the performance of the boys and the ways are not acceptable.

Officers from 9 Paras came and all these boys went around team without taking permission from one. I got annoyed and punish them for the same.

Wrote a letter for Vishal.

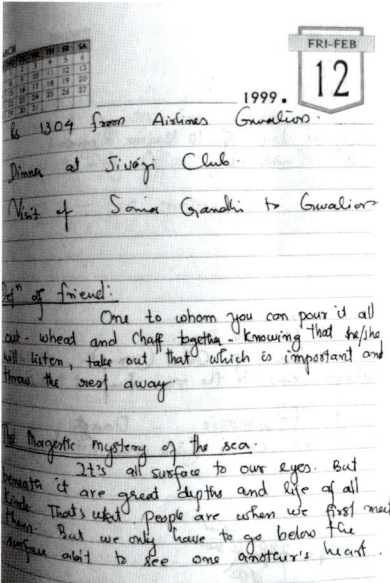

FRI-FEB 12, 1999

Ic 1304 from Airlines Gwalior.

Dinner at Sivaji Club.

Visit of Sonia Gandhi to Gwalior.

Def of friend:
One to whom you can pour it all out- wheat and chaff together - knowing that he/she will listen, take out that which is important and throw the rest away.

The Majestic mystery of the sea:
It's all surface to our eyes. But beneath it are great depths and life of all kinds. That's what people are when we first meet them. But we only have to go below the surface abit to see one another's heart.

SAT-JUN 12, 1999

Battle of Tololing.
started off just around 8:30 PM
and lasted till 5:30 Am.
Casuality — 1 Off
 — 2 JCO
 — 6 OR } own
enemy casuality — 11 kill
Wpn recovered
 — 17 AK 56s
 — 8 HMG
 — 2 RPG
 — 1 RCL
 — 1 DNVD
 — 1 Bino
 — 1 Machine gun tripod
 — Ammo.

13 SUN: assaulting coy
 — D coy led by Maj Mohit S
Reserve — C coy, led by Maj Vivek
 — A coy (Maj P Acharya)
 — B coy (Lt SS Raut)

Gthatath It's not the size of a weapon that matters, but fury of the attack

FRI-JUN 18, 1999

Went out with a party of 2 JCO and
12 OR for search and destroy mission
of enemy mortar position.
It was a fruitless effort.
Had a very hard time climbing the
ridges. We also faced a very difficult
time as there was no water available.
Fortunately we found a piece of ice
which was then melted by heating the
ice pieces in two steel cups with
the help of Napthalien tablets.

FRI-JUN 11, 1999

attack on tololing post poned due to
as the 18 Grn could not
built up their attack on 5140.

and the effects of that, for as long as a year and a half was eating him up every day.

Finally, I made a decision. I went to him one morning while he was just getting ready to start his day. I sat him down, looked right into his anxious eyes and told him, 'Neibu, I know what is troubling you. But, for once in your life, I want you to think about yourself. You have done enough and more for us, and I know you will continue to do so the minute you are back in the position to. I have made a decision. I will take a break from my studies and take up teaching for as long as you are in training. I will make sure that I do enough to earn for everyone until you finish it. No one would have to give up on anything. I will be responsible. I promise you.'

I could see the expression on his face softening while I was talking. When I finished, he told me that he loved me for offering to do it, but he could not expect me to sacrifice my life and dreams for the sake of his. He would not be able to live with that burden. Once again, I put my hands on his, and said, 'I won't be doing that. I will hold the fort with all my might until you are back and start earning your full salary. I will go right back to my life once you are done. This is the best option for all of us. Have faith. We will all get through it.'

Finally, Neibu's face broke into a smile as he hugged and thanked me for offering to step up. He told me that he had faith in me and that now he felt better about taking the step.

The day Neibu was leaving for the academy is as fresh in my mind as yesterday. His bags were packed and ready to be picked up. He had showered and dressed smartly, as he always did, with every crease in place. He had to leave in a few minutes, but he couldn't stop going to the washroom again and again. Each time he would come out with droplets of water shining on his otherwise flushed, red skin. It was the first time that Neibu was going so far from his hometown and his family, and for so long. I knew he was struggling to contain his emotions and was failing miserably. Try as he did, tears were finding their way out of his eyes every few minutes, and he was washing his face over and over to make sure no one else would break down. None of us were used to seeing Neibu like this. I couldn't take it anymore. So, when he came out one last time, ready to leave, I pulled him into a hug. While we pulled away at the end of it, I said, 'Come back soon, my brother. Remember this is temporary. I am waiting for you to come back and relieve me of my duties.'

He laughed and promised to remember that. I believed him.

7
Naga GC Creates History

THE DAY AFTER NEIBU left to begin the rest of his *fauji* (army) life, I went to my college principal to have the conversation that had been playing in my head ever since I had assumed the responsibility of taking care of our family while Neibu was training to become an officer. In a trembling voice, I walked up to her and said, 'Ma'am, I won't be coming to college regularly from now, but could you please still let me write my exams?' Noticing the uncomfortable pause in the conversation that this statement of mine led to, I added, 'My brother is going to the Indian Military Academy (IMA), and I need to take care of my large family while he's gone.'

Confused, she asked, 'What? Why? And what would you do in the meantime?' This was the question that I had asked myself for days before coming up with the answer that would decide the fate of the next year and a half of my life. I answered confidently, 'Ma'am, I will have to teach or pick some other profession to look after my brothers and sisters.'

To my surprise, my principal instantly said, 'I'm proud of you. Go ahead.' Since mine was a private college, a certain percentage of attendance was necessary to be allowed to take exams, but owing to the kindness of my dear principal who accommodated my special needs, I was allowed to write them despite not attending college regularly.

Straight after this conversation, I went and applied for a teaching job at Bethel School in Kohima, where Neibu had worked before he left for his training. I was convinced that this path wouldn't be easy since I had never taught before, and I had been a pretty average student all my life. The only thing working for me in this situation was that I was the most sought-after teacher in Kohima, being the younger brother of Neikezhakuo, whom the principal remembered as a good man. So, when I went and introduced myself as his brother, and asked for a job, the response was, 'Just like your brother, you must be a good man too.' And that's how I earned my first job as a teacher, thanks to my big brother who had a way of being omnipresent in all our lives even while he was away.

That day, once again, my faith in God's plans for us was reaffirmed. As I stepped into the shoes of a teacher, I felt closer to both my elder brothers than ever before. Finally, after all these years, all three of us would have something in common to boast about. Something we could talk about in depth when we sat together.

It's strange, however, how life happens to you in the strangest ways, at the strangest of times. All three of us had gone through most of our childhood and teenage years together as individuals with very different personalities and preferences, and now that I didn't have either of them around, I found myself right in the middle of their world of teaching. Therefore, even though I was not a natural in this role, I was confident that I had it in me to be the best teacher in Kohima—after both my brothers, of course. It ran in my blood, I firmly believed.

I'd be lying if I said I didn't get cold feet before starting out on the job. After all, I was barely nineteen years old. Would the students listen to a young teacher like me? Would they obey my orders the way they did Neibu's? I had big shoes to fill, considering Neibu was teaching them before. I knew I'd be compared with him. His impression in school was that of a 'junior tiger', a moniker given to him in the government higher secondary school. He was a strict teacher.

Our cousin Kiyasetuo, who was also his student, told me once that the assembly hall, which was usually unpleasant to be in because of the students being noisy and undisciplined, would fall into a pin-drop silence when the 'junior tiger' entered the room. 'They were scared not because he would punish them but because, they respected the man of principles so much,' he'd say.

Neibu was not only a teacher; he was an idol for most of his students.

I, on the other hand, was lenient as a teacher; so, even though I was on my way to emulating Neibu, becoming a teacher like him was a far-fetched dream for me. Yet, I tried. Whenever I felt low and nervous, I would think of the advice my uncle Neizieo gave me the day Neibu left for the academy. He said, 'God feeds even sparrows. You are better than sparrows. God will certainly feed you.'

Uncle Neizieo, another mentor in my life, apart from Neibu, didn't only pray for our situation, but he also taught me more ways to make money. He taught me how to make handicrafts and other things I could sell in my free time. I remember how he once challenged me to make a 'bull head' brooch; it took me a whole day to make it. The moment I finished making it, I ran to my uncle and said, 'See, I can do it, just like you.' He said, 'This is not enough, you'll have to keep practising or you will forget.' He also promised to buy the first piece I created to keep it as a souvenir at home.

Back then, he used to sell every piece we made for ₹30. That was motivation enough for me to perfect this art, and I made many pieces with Uncle Neizieo which became popular in our neighbourhood. We were eventually able to sell these pieces abroad too. I even remember sending a batch to Chicago once. This became an additional source of income for our large family—something I'd have never thought of pursuing had it not been for our uncle.

I was excited to show off my handiwork to Neibu, and perhaps the next time he would visit, I thought, we could take some of these brooches as gifts for his soldier friends. How nice would that make him feel! Honestly, it was a relief to have an alternative source of income, and the act of making these things helped reduce my stress.

So, enthusiastically, at the age of twenty, even though I was an arts student, I started teaching math to class four, five and six students. Luckily, I had learned the art of teaching math from Neibu, and I was grateful that I had shadowed him during all those tuitions he took, using his chalkboard in our little home in Kohima. Teaching math was also my way of living my dream of studying science in college and becoming an engineer—something I could not achieve because of my bad scores in the tenth grade.

Soon, I was not only teaching at the school; I was getting requests day and night for tuitions. I would take math tuitions for students up to class eight and finally, the cash I needed as the sole breadwinner of the family started flowing in. I could still not save much since most of it went towards the basic needs of our fairly large household. However, from what I was making, I would save ₹1,000 for Neibu every month; I intended to give it to him whenever I would see him next. It wasn't much, and it would barely pay for his return tickets to Dimapur, yet it was something.

I didn't know then that Neibu, as a Gentleman Cadet (GC), got a monthly stipend of ₹8,000 during the training—a practice that started in IMA in 1941 with a mere ₹21 being paid to every cadet per month. Neibu's was just the first or the second batch of GCs to receive a handsome amount as a stipend for their basic expenses during training. The amount was credited to all the GCs' bank accounts every month, but most of them—either out of laziness or due to procrastination, or simply for the thrill of saving up for something big at the end of the course—withdrew the money in a lump sum only after the course was over. Not knowing any of this then, I was saving up money for Neibu in the hope that I would give it to him whenever he came home on leave.

I wanted to support his hard work at the academy. And I wanted my brother to have some cash with him for emergencies or for when he wanted to treat himself—on their 'liberty' days. 'Liberties' in IMA parlance refers to authorized free time—usually granted every Sunday and sometimes sanctioned on other days by the platoon commander or any other officer in charge, given you are not on any punishment. You could also earn a liberty if you passed an exam on your first attempt!

Parallelly, I had started to dream of the day Neibu would get commissioned as an officer of the armed forces. I knew he would get gifts for everyone like always. At the same time, I knew what I wanted from him. I had even decided to tell him the next time I saw him—a colour television to get through my boredom on dull days.

When Neibu was in the academy, I would often think about him and our conversations about the series of events that led him to his ultimate dream—of donning the uniform respected by society, enabling our large family to live a better, happier life. After all these years, my brother's plan of turning our fate around would finally be successful. It was over twenty years in the making—this aspirational life that would soon be reality for us because of Neibu. He would be singularly responsible for pulling us out of poverty for good. In hindsight, it was a costly dream we had all signed up for.

While Neibu was in the academy, I often wondered how he was doing: Is he being treated like a star in the IMA just as he was when he had reported to the Service Selection Board (SSB)? Back home, we were clueless about what it took to become a soldier, and none of us had ever stepped out of our comfort zones in Nagaland to explore career opportunities in mainland India either. Now, when Neibu had taken that first step, it made me think: Had our family and historical background of belonging to a land always in conflict with mainland India cut the wings of our hopes and dreams? Neibu was the only trailblazer in our family, and I was glad God chose me to be his younger brother. His audacious life choices had, after all, kept me on my toes since my teenage years.

Neibu was great at writing letters, so when he missed us, his home or his friends, he would write. I, on the other hand, simply held his jokes and memories close to my heart and relived them again and again in my mind when I missed him. The last time he was home, before going to the academy and after passing his SSB exam, he shared a joke that went as follows:

> An aeroplane is coming from America to India. As it is overloaded, the pilot says to the passengers, 'Any three of you should sacrifice your life to save the rest of us.'
>
> One American stands up immediately and says, 'In the name of my country, America, I give my life' and jumps into the ocean.
>
> Next, a Russian stands up and says, 'In the name of my country, Russia, I give my life.' He also jumps off the plane.
>
> After some pin-drop silence in the plane, an Indian gets up and says, 'In the name of my country, India, I push this Pakistani off this plane.'

I laughed loudly at the joke then, without understanding the context or emotion behind Neibu choosing to share it with us. The way he told these jokes, with his big eyes and loud gestures, made it irresistible to laugh at them. But I see, in retrospect, that as the academy days were approaching, Neibu's sense of humour was already adapting to the 'army way of life'.

The day Neibu had written his CDS exam—8 December 1996—was a Sunday, the Christian sabbath. A day of church service for Neibu, for as long as I could remember. I think that gave him the confidence that he would clear the exam, as it was a holy day for him and our whole family.

Those days, the aspiring young soldiers would scour the 'Employment News', the Government of India's weekly job journal, for their roll numbers. Finding their number in it was a small yet proud token of recognition from the Union Public Service Commission (UPSC), a prelude to the gruelling SSB evaluation. However, the appointment letter from the Directorate General of Recruiting (DGR) in RK Puram, New Delhi, had the longest waiting time of three to four months. On top of that, there were always postal delays. Without the letter, even if your roll number was featured in the journal, you were as good as rejected. Till the last minute, cadets would be left biting their nails, asking their neighbours and fellow applicants whether they had received their appointment letters. Unfortunately for Neibu, he didn't have anyone around to ask whether they'd received the letter since he was possibly the only Naga boy to have written the CDS exam in our neighbourhood.

Neibu received his appointment letter barely twenty days before the start date of the SSB evaluation. I can't begin to imagine what seeing his roll number featured in the journal would have made Neibu feel. It is the first thrill any CDS applicant gets on the journey to becoming a gazetted officer of the Indian Army.

Until that day, Neibu had only used national newspapers to gain knowledge and equip himself to pass competitive exams. He had never imagined that he would one day see his name in print in a weekly that was circulated just as widely.

Neibu getting through his CDS and eventually SSB were moments of truth for him and our entire family. Were we ready for the lifelong bouts of separation? Probably not. Had we still made our peace with Neibu's decision? Yes, certainly.

One thing we all knew and were confident about was this: Neibu, with his discipline and sincerity, had prepared for this moment almost for his whole life. This moment of glorious truth where his enviable character traits would get him to shine like no other.

Sometime in 1997, Neibu had landed for his SSB. We almost didn't realize when he went and came back with a joining letter in his hand for the IMA, officially confirming his selection. Back then, unlike today, there were a few SSB centres—in Bangalore (now Bengaluru), Bhopal, Mysore (now Mysuru), Dehradun, Benaras (now Varanasi), Coimbatore and Gandhinagar. Some of these were only for the Air Force and Navy selections. Out of all the centres, Neibu happened to be invited to the so-called 'rejection center' of that time—Allahabad. Legend had it that the jury there was known to be so strict with their selection criteria that barely three out of a batch of 135 appearing for the SSB would get picked for the IMA from there.

SSB has an elaborate curriculum to test the candidates thoroughly. You could say that the regime is strategically designed to not be for the weak-hearted, and some candidates even opt out of it midway through the assessment. It is a good four to five days of rigorous evaluation that tests whether the applicants are fit to don the prestigious uniform of the Indian armed forces. The tests include a combination of QnAs, group discussions, and physical and psychological screening. Every test, and every action, is observed by the board during the time the candidates spend on campus; it is perhaps just a teaser of what one should expect from the IMA training.

When Neibu came back home after the five days, he mentioned their routine at the SSB. It was a routine I could not imagine. Their day used to start at 4 a.m. and go on till late evening. The daily schedule would have come as a shock to any young man except Neibu, who was already used to waking up early since the sun back home rose around 4.30 a.m. He was always an early riser—the kind who would wake up every day at 5 a.m. to study, teach, or help with household chores. While the SSB schedule was exhausting, the planners of the curriculum made sure that candidates got little breaks throughout the day. The regime for the day was planned beautifully, divided between a battery of tests, meals in between, tea breaks and physical activities. There used to be multiple batches giving different tests simultaneously every day.

If you passed all the tests, on the basis of merit and with a bit of luck, your roll number would be called out in front of the batch and the presiding officer would stick a plus sign next to your medical chest numbers (new permanent chest numbers allotted to candidates who are chosen for the medical round, as opposed to the temporary ones allotted during previous rounds) to signal that you had made it to the final stage of evaluation. This medical round would usually take one week. Those found to be ineligible would have to go back home. Candidates rejected for solvable problems, like being overweight/underweight, would be called for a re-check after about a month. This is one of the reasons why the merit list is printed much later by the SSB.

One sees all kinds of men at the SSB, grinding together, shoulder to shoulder. In every batch, there would be macho men who were confident they'd get through, considering their strong physique; there would be second-generation and third-generation to-be officers; there were civilians who were the first in their family to become defence personnel; and then there was Neibu—5 feet 6 inches in height, calm and the only Naga in his batch.

Although Neibu must have been one of the few candidates belonging to a 'disturbed area'—according to Indian government terminology—to clear the medical round and enter the SSB centre, he was made to feel at home from his first day there.

Thanks to his unusual background and his humility, Neibu was the star of the batch. Col Sachin Duseja, a course mate of Neibu's who hails from Dehradun, once told me how Neibu was the centre of attraction when he first stepped into the SSB centre at Allahabad. Everyone wanted to have a word with him even though his Hindi was limited to '*Aa jao*' (come here) and '*Chale jao*' (go there).

Duseja narrated to me how Neibu was a magnet, especially for the young lot of NDA aspirants. A conversation with Neibu after a gruelling day would be a refreshing break for anyone. Neibu, without even asking for it, would receive, in advance, tips for the tests. Young candidates would tell him, 'Tomorrow, they'll ask you this, and they may put you into that command task.'[1]

He would politely consume the pieces of information but would never use it or cheat.

At SSB, the more you get noticed, the more brownie points you receive, especially when it comes to physical tests. And whenever there is a command task that would need cadets to pick buddies to help with the task, the popularity of some cadets would become obvious. If a candidate appeared again and again in different tasks, it would imply that he was one of the most trusted in his batch. And Neibu was one of those who did.

The problem statements assigned often look like this:

> There is a well you need to draw water from, and there is no pulley system. However, there is a beam, a rope and a bucket lying around. What would you do?

Usually, for such command tasks, candidates would, mischievously, go on rounds at night requesting the others to pick them on the next day. This was their way of getting an edge over their competitors, but the practice wasn't exactly by the book. Hence, each time someone asked Neibu to invite them for command tasks, he would say, 'No, I am not interested, I will do what I have to do.' Neibu was still called on by many people, considering his physical strength and the background he came from. After all, Naga soldiers are known for their exceptional bravery, strong sense of loyalty and unparalleled combat skills, making them formidable warriors. I feel that sometimes, Neibu's honesty and sincerity got the better of him. Nobody, however, took offence with his refusal to participate in such mischief. The innocence in his smile made up for his curtness on most occasions.

Had I not known about Neibu's experience at the SSB, I would have been worried about him moving to a completely new territory in India, far from our comfort zone.

Impressing the board in Allahabad was no easy feat, Neibu once told me. After the lengthy evaluation process, at the end of the fifth day, the last leg on the SSB journey, when Neibu was called for medicals, the other candidates who had passed with him broke into hearty laughter, filled as they were with relief. But Neibu, shockingly, announced: 'I'm not going to join.'

The fact was that the good news had hit him like a cannonball. I think that was the first time he questioned his decision to join the army. He knew the trade-offs that came with the job. (The second time was right before leaving for the IMA.)

Back home, he had his family and the love of his life, whom he wanted to marry very soon. Joining the army meant giving up on his responsibilities towards the family as well as his better half for the long spells of time that he would have to be away. For Neibu, it was not an easy decision to make.

The dilemma was real—to serve a country that didn't feel like home to many of his fellow Nagas, in exchange for a bright future for his family, or to stay back home with his family and be okay with their current standard of living.

Amidst the volcano that was building up in Neibu's head, all the candidates who had made it with him, including GC Sachin Duseja, tried to cheer him up. They told him, 'If you do this, you will be a game-changer in your family. You have so many brothers and sisters, and this is how you will learn, and this is how you'll get a reputation in your state. And you're from Nagaland, where there are hardly any takers and hardly anybody is stepping into the mainland.'

His answer to all the excitement remained: 'No, I'm not joining.' It was shocking for all the candidates around.

He came back home with that feeling.

To think that Neibu, one of the three candidates to have made it as a gentleman cadet, a badge of honour the whole state of Nagaland still holds close to their hearts, had initially—albeit very briefly—rejected the title in his head and in front of his fellow candidates because of his fear about how the family would survive without him! However, after he and I agreed to exchange family duties for the eighteen months when he would be training, Neibu's excitement of becoming a soldier spruced back up.

In conclusion, around the end of June 1997, he packed his bags to join the most prestigious institution in the country—as a part of the 103rd regular course of the Indian Military Academy in Dehradun—a place even farther away from home.

His selection in the IMA, against all odds, was indeed a testament to his resilience and dedication.

The Indian Military Academy (IMA) marked a significant chapter in Neibu's life—a very different one from the life he had lived up until now. From his first counselling session with Col Robertson and Col J.K.G. Kutty, the first-ever army men Neibu had encountered and interacted with in his life, to being surrounded by only men in uniform, my brave brother Neibu had come a long way. However, the goal in his head remained the same: 'I will change the fate of my family.'

He was on his way to rewriting history for the whole Kengurüse family, and there was no looking back from here.

Although the joining date was 1 July 1997, many gentleman cadets arrived at the plush IMA campus, spread across 1,400 acres, only between 10 and 15 July for various reasons. Some had to give medicals again, some passed medicals later, some were rejected after multiple rounds of medicals, and others received joining instructions late.

Neibu, however, joined the academy on day one. The first day in the academy was mainly spent on academy darshan (tour), the time to get everything together to prepare yourself for the next eighteen months of your life. You pick up your bicycle, pitthu (backpack), uniform for all occasions, shoes and stacks of books. A senior takes you around the academy, gets you familiar with classes, the canteen, the different grounds for different activities and, most importantly, the rules and the punishments for breaking those rules. All this while, you're making mental notes, of especially what not to do while at the academy, because you know that breaking rules comes with a huge cost, not only for the rule-breaker but for the whole batch. That's how the 'buddy system' is first instilled deeply in the hearts and minds of every cadet.

The first couple of days in the academy offered a much-needed lull before the storm that IMA brought with its rigorous training programme.

While we were back home and Neibu was out in the world, creating a place for himself, where everyone looked different from him and spoke other languages, I used to think about him a lot. I used to put myself in his shoes and would try to imagine how lonely I would feel in his situation. At the same time, I was confident that he'd find a way to carve a path for himself in even the most alien of environments. To say that he was a survivor in the big bad world is an understatement.

While I was concerned about Neibu, he was doing more than fine in his new army life. On his second day of joining IMA itself, Neibu had his first encounter with a fellow GC, thanks to the lack of signs or directions when you entered the grand gate of the Indian Military Academy back then.

GC Vishal Singh Bais had reached the academy a day late and after going around in circles, exhausted and clueless about where to report, he had knocked at the door of the first barrack (a house that has cabins in it for the cadets) he saw. GC Vishal was surprised to see a mature-looking man in a nightgown (a traditional-style nightsuit for Naga men) with his hands in his pockets there. Neibu was possibly the oldest guy in his course; he had seen the world, studied and done multiple jobs as a teacher before he decided to take the competitive exams.

After a brief pause, GC Vishal asked Neibu, 'I've come for this course, and I need to report. Can you guide me to the reception?' Neibu instantaneously pointed towards the reporting centre and gave the young man directions. GC Vishal, like any other cadet in the academy, found Neibu's accent peculiar and fascinating. With that first impression in his mind, he reported at the reception and submitted his documents.

As luck would have it, out of the twelve battalions in the IMA, GC Vishal was put in Sangro battalion, the same as Neibu's, which meant he would stay in the same barrack as the one he had landed up at coincidentally when he first set foot on the IMA grounds.

I needed some context when I first came across this tale. I learnt that three platoons make a Company, and two or more Companies make up a battalion. Neibu was a part of the Thimayya battalion, named after General Kodandera Subayya Thimayya. Under the Thimayya battalion, there were three Companies—Alamein, Sangro and Meiktila. In IMA, platoons one, two and three belong to the Alamein Company. Four, five and six made up the Sangro Company. Seven, eight and nine composed the Meiktila Company.

These three Companies (Coys) were the only ones that had been named by the British, after the various battles fought by Indians. Other battalions were all named later, by the Indian Army, after former chiefs of the Indian Army.

Every Company has traditions that cadets are asked to carry forward by their seniors. Like one Company would be the best in battle obstacles, another would always come first in cross-country runs. The Sangro Company, however, had a running joke, considering they were never among the top three in the exercise competitions, that their tradition is to always have 'double figures'.

Although Neibu and Vishal were in two different platoons—number 4 and number 6—both were in the same squad and had chosen each other as buddies, which meant they were required to be seen together in the academy at all times. The rule in the academy is that the juniors can only move around on their bicycles while in a squad, which is made up of no less than four people. So, in case a senior happens to see a GC without their squad, the juniors would have to immediately stop cycling, get down and walk, carrying their cycles on their shoulders. And walking in general was not allowed, unless on medical grounds. If you walked by mistake, you could expect to hear the loud voice of some officer in charge asking you to 'double up'. Therefore, thanks to this regime, Neibu and Vishal were destined to be close friends. In no time, they started to spend all their 'liberties' together, since even though they were not the brightest students in their class, they always passed their exams in the first attempt. GC Vishal would, more often than not, take Neibu to a family friend's place in Dehradun for a home-cooked meal, and Neibu would get his friend Ravannuo to write letters to

Vishal. Perhaps, he wanted them, two of his closest friends in the world, to get along and become closer. Or maybe he still had hopes of Ravannuo joining the Indian Army.

Life and its many facets are not under your control. Things work in mysterious ways, when you're least expecting them to work out. We all know by now that Neibu was not the most extroverted guy, but thanks to where he came from and his name, he had to make no effort to break the ice with other gentleman cadets. That's how he made many acquaintances on his first day when he went to the JQM (Jemadar Quarter Master) store to collect things for himself.

While standing in line, feeling on top of the world despite getting issued the same things as everyone else, some fellow cadet or other would say hi to Neibu and introduce himself. Neibu would invariably reply with pride, 'I'm Gentleman Cadet Neikezhakuo Kengurüse.' After this, the cadets would end up looking at each other, giggling and perhaps thinking to themselves, 'What kind of a name is this?' But before the cadets could walk away, not being able to pronounce his name, Neibu would smile and add, 'You can call me Neibu!' In fact, he preferred cadets and drill *ustads* (those responsible for instilling strict discipline in the boys waiting to become officers; drills were considered the bedrock of discipline) using nicknames like Neibu, Naga GC, Kangaroo, Kengurüse or Neike, as that saved him the horror of his perfectly thought-out name, given by our parents, being spoilt in translation.

Nimbu Saab

In the academy, Neibu found himself amidst a melting pot of cultures and backgrounds, with each cadet having their own personal goals and stories behind their decision to join the Indian Army. When everyone would meet for the first time in the initial days and have casual conversations, they would find some or the other commonality with each other. They would call it 'types' if two cadets found something in common between them. There were many, many types of people in the course. Place types, college types, SSB types ... but there was no one like 'Naga GC'—a nickname coined by drill instructors for Neibu, who was one of a kind in his batch. Thankfully, Neibu ended up making some friend or the other among all 'types'. Along with his closest buddy GC Vishal, there were: GC Praveen Tomar, a third-generation officer and ex-NDA; GC Sachin Duseja, famously known as 'local GC' since he belonged to Dehradun who, like Neibu, would have been a first-generation officer; and GC Sundeep Khatri, an ex-NDA and a long-distance runner, who eventually turned out to be a great teammate for Neibu in football and obstacle courses.

There were many more cadets Neibu would often mention in his letters. Reading about them used to give me some relief. Wherever he was, he was not alone.

While some of his friends were his football teammates, others were his battalion type. With all his enthusiasm for learning again, Neibu made sure he fit right into the system.

When ex-NDAs joined in the second term, they walked into the institution boisterously, like they owned the universe and nobody could beat them at IMA, but they had nothing on Neibu when it came to his fitness and leadership attitude. It wasn't Neibu's age that gave him the wisdom he needed, it was his experience in the real world that made him who he was.

Neibu, by the second term, was not only playing sports like never before, he would also volunteer to take on more tasks, not to impress anybody, but because he simply was one of the oldest yet fittest guys in his course—even though he was a 'direct entry' and not ex-NDA. Owing to this competition with Neibu, sometimes, ex-NDAs, in their childish way, would make fun of Neibu's name and accent. In fact, some time in the third term, a couple of ex-NDAs even got into a hand-to-hand fight with Neibu. GC Vishal had to step in to resolve the matter and also received a few punches in return, unfortunately.

Unlike most ex-NDAs though, GC Praveen Tomar shared a camaraderie with Neibu which was different and like a kind of brotherhood. One incident that solidified their friendship occurred in the third term, during camping. In camps, cadets try their best to take the easier path to the finish, with a lighter load, and mostly, ex-NDAs seem stronger to shoulder more weight in this activity. On this occasion, the task assigned was to carry weapons from point A to point B. Nobody wanted to take the Light Machine Gun (LMG) assigned because it is heavy—double the weight of other weapons. So, during this particular drill, when GC Praveen Tomar asked, 'Who will carry the LMG with me?' Neibu raised his hand promptly and shared the responsibility with him. They alternately carried the weapon till its destination.

With every incident and drill like this one, Neibu's relationship with his new-found friends became stronger. In the true sense of the word, these soldiers would soon be brother officers.

Yet again, like any other place Neibu had lived in, he'd thus found family in cadets like GC Vishal Singh and GC Sachin Duseja, both of whom had some family members living in Dehradun—Duseja had his birth family, and Vishal had his first cousin. That year in the IMA, there was a saying in vogue that went like this, 'Never stand next to Naga GC or a Sardarji. You'll get caught!' The basis of this saying was that since both Sardarjis and Nagas looked different from most others around and could be recognized from afar, the cadets standing next to them were sure to get caught and be punished. Still, this didn't stop Neibu's friends from taking him with them on the rare occasions whenever he'd say 'Yes' to go out with them for fun.

The gentleman cadets would also jump at every opportunity to take Neibu home to give him a taste of their home-cooked meals. Now, unlike liberty days, which was the term used for authorized leave for gentleman cadets, there also existed 'french leaves'. French Leaves (FLs) were the leaves gentleman cadets took illegally when no one was looking. They came at a price—the risk of punishment from the authorities if they got caught. It was a risk that cadets would sometimes take when surveillance in the academy was low.

Neibu was not a law-breaker, and there was only one time that he took this risk. Neibu was in his third term, and probably in a bad mood, when GC Sachin Duseja asked him to go for an

FL with him, to his house. He said, 'Let's go, we'll have a home-cooked meal. There are no scanners and so, it's very safe.'

GC Sachin was certain he would get a 'No' in response, but to his surprise, that day, Neibu said 'Yes'. Until that day, Neibu had spent most of his liberty days in his room, or with GC Vishal. But now, the term was almost coming to an end, so I guess that's a call Neibu took to experience something new.

Neibu and GC Sachin, thus, took their bicycles and left the academy for their day of fun. If they were on authorized leave, they would have had to leave their bikes at the gate, show their 'out passes' and then take local conveyance to wherever they wanted to go. Since that was not the case this time, they headed out on their cycles. On the way, Neibu stopped to buy some things from a shop, as this was a rare occasion when he was out and could visit a store that was not an army canteen.

Both the gentleman cadets got off their bicycles to enter the shop, and the unexpected happened. The academy adjutant, Lt Col Amre, was standing right in front of them. A man who had a bone-chilling effect on cadets. Duseja could not believe his eyes, and he could not believe that the one time he got caught in the academy like this, it was with the 'Naga GC' who would never cover up his tracks with an innocent lie or an excuse. They thus found themselves standing right next to the man the whole academy was advised not to cross.

Lt Col Amre asked them, 'Gentlemen, what brings you to this shop?' As expected, Neibu surrendered immediately.

To this, Lt Col Amre replied, 'Go back and tell your Company commander that Lt Col Amre wants to speak with him.'

The poor men had sneaked out with many aspirations and were caught in the first mile itself. To this, Neibu said, 'I knew it, I'd committed a sin; God punished me.' To lighten the mood, GC Sachin Duseja jokingly responded, 'Father, if you knew so much, you should've advised me earlier.'

With many other fond memories like this, Neibu inched closer to his graduation day.

The IMA transforms a young man into a determined soldier. It prepares the young cadets to lead well—whether in battle or during peacetime. The schedule and the environment in the academy can turn the most undisciplined man into a solid soldier, and Neibu was much ahead in this race of becoming a gentleman officer. The early mornings, the rigorous training sessions and the constant push to excel became Neibu's new norm.

Sometimes, I thought how funny life was. After all these years of being a teacher, Neibu had gone back to being a learner, a student.

In the academy, it's often said that 'You should be at the right place, at the right time, in the right rig (clothes/uniform)', and if any gentleman cadet missed this memorandum of conduct, the whole batch would be punished—together. The only times such punishments were imposed on Neibu's batch on account of him, it was primarily because of his outstanding personality traits—especially his honesty. He would never lie to the authorities when he made a mistake or could not complete a task.

For instance, this one time during the second term, the term when NDA graduates also join the academy, a senior called everyone to 'fall in' and asked the boys to do fifty push-ups. Initially, Neibu used to find it difficult to even do five. While most gentleman cadets on this day cheated and got done with the task, our dutiful Neibu tried his best to do all fifty, and when he failed, he honestly told the senior that he couldn't do it. The whole batch thus had to complete another punishment with him. That day, the whole batch probably would have been angry with Neibu for not faking it, but this sincerity of Neibu's made him one of the strongest cadets by the time he passed out of the academy.

He never left a task unfinished and gave his best to each one that came his way. In an exercise where buddies were supposed to pick up their respective partners one by one and run as if one were in a war scenario, Vishal used to have a ball sitting on Neibu's shoulders, while he really struggled when Neibu would sit on his. He would exclaim, 'You look so tiny but are so heavy!' So even though when he joined the academy he was average in physical activities; by the time he left, Neibu could beat anyone in the same tasks he found difficult to do at first. His growth trajectory in the eighteen months of being a GC was the most impressive out of all the cadets.

As Neibu's graduation was approaching, Neibu and his friends were training hard, day and night, to pass out of the IMA with flying colours. During this time came a record-breaking moment, wherein the Sangro Company came third in the last drill competition of the course. Their drill ustad was left stumped, and the Sangro Company boys jocularly feigned a sigh, as if to say, 'Oh we broke our tradition!'

Meanwhile, I could not help but be a little selfish and feel happy that I would soon be able to pass my duties back to the man of our house. The only man who could do justice to these family duties.

I was exhausted yet excited for the D-day and for the rest of our lives.

8

The Rise of Lt Neikezhakuo Kengurüse

Perhaps very few times in my life have I felt the excitement that I did when we were gearing up to go for Neibu's Passing Out Parade, or POP, as they call it. Not only were we going to see Neibu after a long time, but we were also finally going to see him in uniform! The first army officer of our family. The very first person who had shown the courage and determination to break the boundaries of what had seemed to be our family's destiny for as long as we had known. He had not only dreamt it; he had lived it.

Years later, I found out how Neibu had made the decision to select an arm from the services—once again, he had done it based on the needs of the family. Since he knew how much we would need him, he had picked the Army Service Corps, or ASC. He had somehow gathered that the postings in this corps were relatively better than others. As he already belonged to a hard-to-reach corner of the country and wanted his trips back home to be convenient, he opted for the ASC. Luckily, he also got what he had opted for. However, as was the practice for all gentleman cadets joining the services as an attachment, he first had to serve in the infantry for a period of two years upon getting commissioned. For that purpose, Neibu was going to get commissioned in a very illustrious unit—the 2 Rajputana Rifles (or 2 Raj Rif). Incidentally, two more of his course mates and fellow ASC officers were chosen to join the same unit—GC Vijyant Thapar (also known as Robin) and GC Praveen Tomar.

Nimbu Saab

Since none of us had stepped out of Nagaland and that too to a destination as far as Dehradun, our cousin Neitho graciously offered to come along with us for Neibu's POP. After an uneventful, rather long journey, as the train screeched while slowly pulling up into the Dehradun station on a cool winter afternoon, our anxious eyes couldn't stop looking out of the window, trying to spot the familiar face we longed to see amongst the young men around, who were all similarly dressed, in muftis, and were trying to scan coach after coach for the faces of their own loved ones. As the train came to a stop, Neitho and I huddled around my father, making sure he got off safely while our shoulders supported the weight of our modest bags, which were carrying our clothes for the next two days.

We had barely collected our bearings on the platform when we finally spotted the familiar, beaming Naga face we were looking for. Neibu, who was walking towards us, was dressed smartly in what appeared to be a grey suit that showed off how fit he was. It was a snug fit but was not tight. Inside, he had on a neatly tucked white shirt and a blood red and steel grey tie, which he told us later was supposed to signify blood and steel. Though Neibu had always been a smart dresser, there was something different about how he looked now. I realized it wasn't just his clothes, but also his body language. He seemed to be carrying himself with so much confidence. How handsome he looked. For a minute, it was hard to imagine it was the same simple but hard taskmaster we all knew and loved back home. Now he was an elite officer!

The Rise of Lt Neikezhakuo Kengurüse

He gave each of us a warm hug, squeezing me and Neitho a little tighter in the end. I saw later that it was a common way of hugging for all the gentleman cadets in the academy. Some even lifted the person in the end before letting go.

From the station, Neibu took us to another Naga boy's house. This boy, Kuolie Mere, was doing his post-graduation in Dehradun and was someone we knew from back home. Neibu knew that we would be comfortable staying with someone familiar, as coming all this way to a new city amongst all these new people would be overwhelming for us, especially for our father. Traditionally, families of other gentleman cadets were accommodated in different barracks, while gentleman cadets had to make space and move in with each other in twos or threes. (During this time, GC Tomar was sent to Neibu's room, where he learnt about his past as a teacher.) But Neibu had chosen for us to stay where he thought we would be more comfortable. After spending some time with us at Kuolie's place, Neibu took our leave, letting us know that he would come back the next day to take us around his academy, the IMA.

The moment he arrived back at the campus, he also received his friend Vishal's parents and enthusiastically took them to meet their son. His excitement knew no bounds when he knocked on GC Vishal's doors. It all seemed like one big celebration to him.

As promised, Neibu was back rather early the next day and waited patiently while we got ready, one by one, in the small, rented apartment. I found myself looking at Neibu's face from time to time. He had really done it. He had gone ahead and had paved the way for all of us. Our future was going to change, and we all had Neibu to thank for it. Several times in the day, I would find myself overwhelmed by the reality of it all. It was like one of those things that you think will never happen for you, until it does.

Finally, we were all ready and on our way to the IMA. Neibu kept pointing to little nooks and corners of the city, explaining their significance, until we reached a road which seemed to be away from the buzz of the rest of the city. A little further down the road, we could see neat boundary walls lining both sides of the road, shielding the wide expanse of the academy inside from the main road. In no time, we found ourselves on the other side of the huge, daunting, heavy iron gates that opened to a narrow path with small lush green lawns on one side and a massive, almost palatial, structure on the other, the Chetwode Building.

The whole day we went around the huge academy as Neibu showed us his barracks, the different courts and grounds where they did their drills and other such places within it, finally ending our tour at the canteen, where we sat down for lunch. All the time I kept marveling at how neat, clean and organized this place looked despite being so huge. I loved to watch other junior GCs go ramrod straight while passing senior GCs and officers. The discipline in the place seemed to be impeccable. I felt so proud knowing that my brother had been a part of, and was now passing out of, such a prestigious institution. I was definitely awestruck.

The actual passing out parade the next day surpassed all my expectations. Though none of us knew what to expect since this was the first time anyone in our family was joining the army, we had still held on to our expectations, based on whatever we had watched on television or heard from people. But the real ceremony

The Rise of Lt Neikezhakuo Kengurüse

went far beyond whatever it was that we were hoping to watch. Several times during the ceremony I could feel goosebumps rise on the surface of my skin, and the cold breeze of the winter morning wasn't the reason for their appearance. Perhaps it was the first time that some part of me really felt we belonged—that we were Indians.

My favourite part, however, was the end. I thoroughly enjoyed witnessing it. It seemed that all the strict discipline and decorum of the day seemed to melt into one big disarray of celebration, as if in perfect contrast to the rest of the day. Just as the ceremony commenced and the oath was taken, all the GCs threw their hats up in the air, screaming at the top of their voices, hugging each other, while some groups did activities like push-ups or some kind of jumps together with those from their Companies, elation written large on their faces!

Finally, it was time for the pipping ceremony, where the parents of GCs were supposed to remove the cloth covering their pips. This would mark the last step for them before they became commissioned officers of the Indian Army. Neibu stood proudly in front of our father as he carefully removed the cover from his shoulders. Once the pips were uncovered, Neibu knelt in front of Apuo and asked him to pray for him. A pastor for most of his life, Apuo was used to praying for people seeking it from him. Therefore, when he kept his hand on Neibu's head and found himself struggling to go through because of an uncomfortable pain in his heart, it sent a chill down his spine. This had never happened to him before. Neibu, who had grasped his dilemma almost immediately, said nothing and instead sat still, praying for himself before rising back up with a warm smile, as if to tell Apuo

Nimbu Saab

that it was okay. While I had seen it all, I didn't know what had really happened until I asked Apuo to recount his memories from that day for this book. Is it possible that Apuo had a premonition? I am not sure, but when I think about it now, about how when it was time for the one who looked out for all of us to receive some prayers for himself, he couldn't. It appears that my father and Neibu had sensed something that day.

Meanwhile, all I was looking forward to after all of it was over was going back home. We were supposed to take the train back home, which included a break journey in Delhi. I was all set to finally ask Neibu for what he owed me. I had kept my side of the bargain; it was his turn now. I had my eyes on a colour television set, which was all the craze back then and was easily available in Delhi. (Back home, there was a waiting period for this product.) I had borrowed ₹20,000 from an uncle before our trip, and when I saw that there was ₹5,000 left over from this, I decided to repurpose it. I would only have to ask Neibu for another ₹10,000 in order to buy the TV. I almost couldn't wait to carry my new television set back home and show it to my brothers and sisters.

While our understanding was that I would go back to my studies once Neibu started earning a regular salary, by this time I had been able to secure a government job as well. So now a TV was all I wanted. With a secure job and no one else to look after other than myself, my life was set, and the TV was all I needed to complete the picture. I could earn, spend and enjoy my own life.

Little did I know, I was in for a shock. What I could see as a perfectly reasonable request, especially since I had rightfully earned it, my elder brother saw as a frivolous one. He turned the request down, politely but firmly. No questions asked, no arguments entertained. My pride wouldn't let me ask again anyway, so the matter ended with the 'no'. If he could find it in himself to deny me something that he could see I really wanted, then I too could find it in myself to accept it. Maybe being out and about, away from home and living in a bigger city, and becoming an officer really had changed him after all, I thought. Disappointed and disgruntled, I chose to keep my distance from him and didn't talk much for the rest of our journey.

Back home, everyone was elated to meet Neibu. There definitely was something different about him now, I would think often. He seemed to be happier, less burdened, and he walked around with such confidence. We could also see him focus a lot more on his physical health. He wouldn't ever miss his morning jog followed by exercise.

This was perhaps the first time that Neibu was in Nagaland and did not have to be constantly at work. He was truly on a vacation and was clearly making the most of it. He would spend most of his time outdoors, visiting our extended family or friends. We would often hear him correcting people when they addressed him now, asking them to include his official rank, which he had earned with such hard work. My father's fears of whether he would be accepted by the community after joining the Indian Army also turned out to be largely unfounded, because most people not only accepted him, but they also did it with pride. They were proud to have him amongst us. His friends Ravannuo and Sanuo, who were excited to meet him after such a long gap, would make him dress up in his academy tracksuit and hold him from either side while walking up to the uniformed guards near the cantonment area in Kohima, as if to show off their officer friend! This was also the time that he brought Caroline home, albeit just as a friend, perhaps to see how the family liked her, which everyone truly did.

I couldn't help but notice how Neibu seemed truly happy and carefree for the first time in his life. Sometimes I did want to spend more time with him and accompany him everywhere he went to visit people, but my anger towards him stopped me from doing so. It got worse when I found out that when one of our sisters had asked him for some money for a dress, he had been quick to give it to her. I couldn't understand how he hadn't seen that as frivolous, but my TV set, which I needed to have for some recreation in my otherwise boring life in the village, was deemed unnecessary. How had he forgotten all that I had done for him so easily? Did he not see how unfair he was being to me?

The month of December, the time of winter vacations and Christmas celebrations, often seems to fly by too quickly in our part of the country. The December of 1998 also seemed to whizz past us, and it was soon time for Neibu to go back and join his unit, which was stationed in Gwalior then.

To get there, Neibu had to take a flight, and flights were not as common as they are right now. As luck would have it, due to the weather conditions at the time, Neibu's flight got deferred. On the other side of the map, his other two course mates, Praveen Tomar and Vijyant Thapar had reached the unit, one after the other.

The Rise of Lt Neikezhakuo Kenguruse

Tomar, who was ex-NDA and hence more proactive, reached a day prior to the joining date, while Thapar arrived right on time. The unit had to wait for the arrival of their third young officer, Neibu, who they thought was absconding! Since there were no mobile phones in those days, placing a call to army numbers was an issue, and so Neibu hadn't been able to inform anyone about his whereabouts. When an entire day passed and the Naga officer had still failed to show up, the people in the unit wondered if he really was an officer. Could it be that he was an undercover spy?

Neibu finally reached a day and a half late. The delay worked well for him since his predecessors had already borne the brunt of the pranks that are played on young officers as part of their welcome. Since no one even knew when Neibu would be arriving, he was spared. He simply reached his location, met with the officers, was assigned to his Company, the Delta Company. Tomar and Thapar had already been assigned to the Charlie and Alpha Companies, respectively. Incidentally, Tomar and Neibu belonged to the same Company in the IMA, and during the POP, due to the shifting of rooms to accommodate parent guests, had become roommates for a brief time. Hence Neibu had a familiar face in him when he reached the unit.

While the single officers usually stay in rooms in the mess, in a unit, the men stay together in a dorm-like place known as 'lines'. However, for the purpose of training and familiarizing newly commissioned officers to the most basic level of work in the unit when they join, they are made to stay in the lines along with the men of their Company. It also helps them break the ice and build camaraderie with the people they may have to go to war with some day. By the time Neibu had reached the unit, Tomar and Thapar were already staying in the lines and Neibu joined them. The distribution of Companies worked out very well for Neibu as well, since the one he was assigned to was made up of troops (Rajputs) who would eat meat, as opposed to the Jat troops, who didn't.

In any case, Neibu did not really face any issues adjusting to his new surroundings. Those who believe in destiny may think that growing up in a large family living in a small house was perhaps God's way of preparing him for his destiny—for his days in the army. In fact, many of the other men too came from backgrounds similar to our Neibu's, so relating to them was easy for Neibu. However, there was one hiccup. Though a year and a half of academy training had made Neibu comfortable with conversing in Hindi, none of the officers and men had any experience with the Naga language. Therefore, just as it had been in the academy, most people struggled with his first name, Neikezhakuo. So, while the officers settled for Neibu, which was much easier to pronounce, the men gave him a name that rolled off their tongues much more easily. And that is how Neibu became the much-loved Nimbu Saab!

The life of a newly commissioned officer in the unit is rigorous, to say the least. If anyone had ever thought that after the hard and tiring life of the academy, he would enjoy a better one as a commissioned officer, he would be in for a rude shock.

Mornings began as early as 5 a.m. in the lines, and even in the cold months of December and January, there was no geyser or water heater, so everyone had to take a bath with cold water. That was the downside of getting commissioned in the winters as opposed to summers. The routine was the same for everyone living

The Rise of Lt Neikezhakuo Kengurüse

in the lines. After the cold-water bath, they all had to rush for 'muster' fall in, followed by PT. Once the PT was over, depending on whether any of them were called aside by the adjutant (an appointment in the unit) or the CHM (a Company Havaldar Major), they would head back to the lines by 7 a.m. Then again, they had to change, get dressed, go to the langar (army kitchen) to eat and leave to be on parade duty by 8.15 or 8.30 a.m.

They broke off for lunch around 1 or 1.30 p.m., went to the langar to eat and back to the lines to rest for the little time they could manage. The rest period usually didn't last for long since they were required to change and turn up for evening games again at 4 or 4.30 p.m. (The weather played spoilsport in the winter, when days were shorter, so the games were at 4 p.m. in wintertime.) The games would conclude by 5 p.m. or so, and again the young officers needed to be back at the office to complete various tasks allotted to them till around 7 p.m. However, for at least one of them, the day would still not be over as they were assigned guard duty, on a rotational basis, to cover the late hours of the night.

Adding to all the above, at the time of their joining, the unit was preparing for a cross-country run, and all three new officers were put in the cross-country team. This added more time to their physical training since, in essence, now they were going for a 10-kilometre run during both halves of the day.

However, physical training was the least of Neibu's worries, since the unit was gearing up for one of the most-awaited events to take place every time a new officer joins it: the dining in.

Typically, the dining in is supposed to be a formal dinner, welcoming the new 'baby of the regiment', where he gets a chance to meet all the officers and their families, who are also invited for the dining in, to the mess. While the idea is to break the ice for and with the new officer joining the unit, a dining in usually involves much more for the young officers in question. Usually, different units have different dining in traditions, but the one thing that is common to most is the punch that they prepare and that is to be had by the new officer in one go.[1] The punch drinking typically has the desired effect of getting the inhibitions of the new officer lowered, enabling everyone to indulge in a carefree manner. The ritual often leads to stories that are fondly recounted for years and years to come, often embarrassing the person in question.

The dining in organized for Neibu, Tomar and Thapar was no exception. The fact that the three of them had varying alcohol intake thresholds made the evening even more interesting. Neibu and Thapar being complete teetotalers were the first to get completely inebriated. Neibu had never even tasted alcohol before, and so he really was the first one to get sloshed, followed soon enough by Thapar. Within the next few minutes, both of them were found sitting in the dried-up drain outside the officers' mess, cursing the rest of the officers for getting them drunk, saying, 'Jesus Christ will not forgive you, for you have made us sin!' Tomar, on the other hand, was a seasoned player and could hold his drinks better. As a result, the evening ended with the mess running out of the alcohol and Tomar helping both his course mates get back to their beds by 3.30 a.m.!

The next morning, the youngsters were supposed to be on parade at 5.30 a.m., and once again, Tomar, who was an ex-NDA student and thus more hardwired to both alcohol and routine,

The Rise of Lt Neikezhakuo Kengurüse

turned up on time while the other two nursed their hangovers. When they met him later, Thapar accused him of being a *gaddar* (traitor) for turning up on time, leaving the other two behind. Tomar smiled, told them he only realized they weren't there upon reaching the ground and playfully apologized, stating that he couldn't have helped it anyway as he was a creature of habit.

By now the three course mates had completed their time in the lines. It had worked very well for Neibu since, despite the language barrier, the troops had been able to spend some much-needed time with him and interact with their Nimbu Saab in close quarters. They lived, did their jobs and ate together during this time. Through some of their conversations they realized that Neibu too came from a humble background, just like some of them did, and in fact had been a teacher before getting commissioned. Some also heard stories of his large family and many siblings. Their common ground earned him their acceptance. They realized that he was a simple man but was very particular about his work, and, just like he was with his siblings, he would motivate his fellow soldiers to keep aiming higher in life, often asking about their families and encouraging them to make sure they were taking care of their siblings and that they were all educated too, like he had done.

That their Nimbu Saab came from a remote corner of the country, had a humble beginning and belonged to a large family with limited means, and yet had done enough to manage to become an officer, really made them respect him for all he had achieved single-handedly. His undying spirit to do more moved them. In no time, he was one of them. They looked out for him in quiet ways—for instance by saving him some meat on days when they knew that meat would not be prepared in the langar for religious or other reasons.

By the time the three newly minted officers had arrived, the unit was also gearing up to move. Their time in Gwalior was over, and they had to move to the Valley (J&K). Thus, by the time they were moving to the mess (after completing their time in the lines), the mess did not have enough room for all three of them. Since the unit relieving them at Gwalior had started coming in too, the space was limited, and so it was decided that two of them would share one room and one would proceed for leave. Neibu's hometown was the farthest, so him going home was not much of an option. So, amongst Tomar and Thapar, it was decided that Tomar would proceed on leave.

Since the battalion was to move to the Valley soon, the training for counter-insurgency, or CI, operations had started too. The commanding officer, Lt Col Ravindranath, would plan different activities to make sure everyone was working together as a unit. A section (a small group of people) commander's competition was a part of the training, where the officers were given full independence to train their men.

Thus, the last few days in Gwalior were spent in training activities and sports that Neibu loved to be a part of, and the packing up of the entire unit, which in itself was a herculean task. Neibu's letters, though sporadic now, continued, and we could feel his happiness at being in this new environment. He didn't just write to us; he continued writing to his friends too and shared his life's new chapter with them as well. We were all happy that Neibu had found his path, the path that he had dreamt of and worked so hard for. It was inspirational.

Finally the 'special train', a term that was used for military trains carrying battalions on the move, left the station at Gwalior for its destination, Jammu, with the '*veer bhogya vasundhara*' (the motto of the unit, meaning the brave alone will inherit the earth) battalion. Such trains usually came with the comforts of 'home', complete with beds and one's favourite fare on board.

While passing through Delhi, the train had to halt for a few hours, as is usual with these trains since they don't have a schedule to follow like other trains and so, are low on priority of clearance. Thanks to that halt, Neibu had the chance to meet the families of fellow officers like Lt Col Ravindranath, Maj. Vivek Gupta, Capt. Mohit, Capt. Umed Singh and Thapar, as they all got together to cut a big cake brought in by Thapar's parents. The lettering on the cake read: 'Good luck, 2 Raj Rif'. In addition to that, a further delay also gave them a chance to visit the home of a unit officer, Maj. Kala, who was supposed to join them later in the Valley. Maj. Kala and his wife hosted the officers for dinner, and even though he was shy at first, as the evening moved along, Neibu relaxed and opened up, and they all had a much-needed, fun and joyful get-together. The party only finished in the wee hours of the morning, which is when they returned to their special train.

After a rather long journey, the train finally reached Jammu, where they were to halt for a few days before finally moving ahead. Their third musketeer, Tomar, came back from his leave at this point and joined them in Jammu.

Nimbu Saab

While they were there, they found out that Thapar wanted to meet a girl he liked, who was the daughter of an army officer and lived with her family in Jammu. Lo and behold, Thapar, Tomar, Neibu and Maj. Vivek Gupta got into a two-and-a-half-tonne—a mini truck-like military vehicle, also known as *dhai tonne*—and accompanied their fellow officer to help him get his love story started!

In addition to working and training together, it was these small personal things that made officers truly become brothers in arms, and Neibu was glad to be a part of it all. With every passing day, his sense of belonging was growing deeper. His life had a different flavour now, and as always, he was looking forward to the next step. After a few days, they were all packed up to finally travel towards their place of posting ... Kupwara.

9

'Your Own Ease, Comfort and Safety Come Last'

I DON'T KNOW WHEN Neibu started maintaining a journal but I'm so glad he did. It preserved for us the essence of his life, which he lived fiercely, on the principles of duty and honour. It is probably the sole reason I have been able to vividly imagine and talk about his short yet remarkable time in the Indian Army. Through his words, I have truly realized why a uniformed man on the battlefield would readily take a bullet to save his brother in arms. The camaraderie that develops between cadets in the NDA, IMA and beyond is unparalleled—compared to anything I've seen or felt before for someone not related by birth.

The combat uniform of the Indian armed forces had gifted Neibu with a sense of belonging that most of us can only dream of.

After learning about Neibu's military life, the following lines, imprinted on the walls of Chetwode Hall in IMA, Dehradun, made all the more sense to me:

> The safety, honour and welfare of your country come first, always and every time.
>
> The honour, welfare and comfort of the men you command come next.
>
> Your own ease, comfort and safety come last, always and every time.

Nimbu Saab

Over the years, every single time I have had a conversation with a course mate of Neibu's or the men he commanded, these lines have reverberated through my head. It was clear that the time Neibu spent in the academy and after that, in Gwalior and J&K, with his regimental brothers had changed him for the better. He had become more outgoing and extroverted—the only two things that were missing in his life in Nagaland, where I believe the burden of our domestic responsibilities weighed upon him constantly.

After hearing stories from his course mates, of the fun and frolic they had even in the toughest of times and during the most challenging of training sessions, I have realized that during 1997–99, he had become more like family with his men than perhaps what he was even with us at home. He was truly living his best life in the army, even though it unfortunately turned out to be short-lived. It seems as if he had saved his youth to pack up all of life's enjoyment in the two years of living the *fauji* (military) life. He would often step out to go for movies, like *Rush Hour*, *Border*, *Kuch Kuch Hota Hai*, attend dinner parties in the officers' mess, family dinners with friends and even make phone calls on behalf of other officers to their families, to pass on their messages.

I got to know later that it was typical for soldiers posted to peaceful locations to call the families of their brother-soldiers stationed in conflict zones as a favour, since they knew that the latter would have limited means of communication and wouldn't be able to reach their loved ones. During that time, satellite phones and letters were the only means of communication in a conflict zone. Both methods did not provide instant relief for a family in any distress situation.

Before 2 Raj Rif hit their home base, Kupwara, in the period from mid-February to mid-March 1999, the brave men had done their pre-induction training in the Valley, in Khrew, preparing for a war-like scenario. The training went on for a month at the Corps Battle School. This is where another young officer, a junior who had passed out of the Officers Training Academy (OTA) three months after the current lot, Lt Rauthela joined. His dining in, where he was asked to sing *'Nanha Munna Rahi Hoon, Desh ka Sipahi Hu'* (I'm a young man on this journey, I am a soldier of this land) after drinking the aforementioned infamous punch, gave the soldiers some respite in an otherwise disturbed environment that kept everyone on their toes. It is traditional practices like this that make the *fauji* life what it is, on land or at sea, through peaks and valleys, in the quiet of the desert and through the din of the jungle. By the time you become a senior officer, you have been pushed to your limits and are thus able to make the best of any lemons that life gives you.

After crossing Udhampur, Banihal, the transit camp at Srinagar, and after spending some time in Khrew, 2 Raj Rif arrived at the more terror-stricken city of Kupwara along the border of India. As an extension to their CI operations, Tomar, Thapar and Neibu were welcomed in Kupwara with tougher war exercises, such as forty-eight-hour ambushes, seventy-two-hour patrolling, area-domination patrol, and cordon-and-search operations. Anything and everything that would prepare a man in uniform to act strategically in a conflict zone.

For years, I imagined Neibu in Kargil and could relate to the terrain—the altitude of the place as well as the locals residing there, considering their situation seems so similar to what we have faced for years in Nagaland. However, when I finally visited Kargil in 2021 as a guest of the Indian Army to celebrate Kargil Vijay Diwas, two decades after losing my brother to the same merciless peaks, I realized they were way stiffer and higher than my imagination. Neibu's fellow soldiers would tell me later that my brother had a sound knowledge of the terrain and knew exactly what to do to survive there. This surprised me. The mountains there were very different from what we have in Nagaland. The peaks are barren, completely devoid of vegetation. The wind is so strong, yet the lack of oxygen leaves a person restless, fighting to stay calm because of the breathing difficulties one faces there. It is akin to breathing in the air from a vehicle's silencer. You feel the air gushing, but you cannot breathe properly. The houses of the local people and their everyday life is so different from the Nagas even though our stories are so similar in some ways. I noticed that their houses were made of mud. And therefore, unlike the Nagas who are always praying for monsoons to arrive on time, the locals here pray for no rains—because when the rain does come, it often destroys the only homes they have.

As I stood facing the Tololing Top and Three Pimples from a distance, my right nostril bleeding a little due to the lack of oxygen,

'Your Own Ease, Comfort and Safety Come Last'

I found it hard to visualize how a war was fought in this unforgiving environment in 1999. On the other hand, bordering Kashmir are gentle hills, cloaked in pine forests that make it a Jannat (Paradise) that's worth fighting for. That such a breathtaking landscape is perpetually tainted with the ugly marks of terror is really a pity, and it makes me sad beyond words.

Considering its history, filled with destruction and little development, it's no surprise that while there are some prosperous apple traders, the majority of the populace in this state lead a modest life. The livelihoods of the masses often depend on harvesting apples, packing them into crates and loading them onto trucks destined for markets in cities like Chandigarh and Delhi. Tourism (most definitely) is another way they make money, but it's a highly unreliable industry for the Kashmiris, considering the constant trouble in the region. Untouched and serene, the area offers a slice of nature in its purest form—something I'm certain Neibu would have appreciated.

Beneath this calm exterior, however, lay a troubling reality that the Indian Army, unfortunately, had to face daily back then. The ongoing violence and militancy in Kupwara often present a stark contrast to its peaceful facade.

A little before going to war, though, just after going to give his promotional part B exam (if he passed this exam, he would be promoted to the rank of Captain), Neibu visited home again. This was barely three months into getting commissioned.

Nimbu Saab

Our understanding of the army was limited, so Apuo had several questions in his head—like has he run away, been fired, and is it even authorized leave?

The day Neibu landed in Dimapur, the bustling capital of Nagaland, he bought a lot of meat on his way to Jalukie, cooked it for all of us and we enjoyed a hearty meal together. The whole family had a party with him and enjoyed his stories over dinner, just like old times. I wish I had known then that this would be the last time I could ask Neibu about his experience of serving in the army—about where he was based and what their days looked like. But since I was angry with him, I either listened to him talking without showing any enthusiasm or overheard him telling stories to our family and friends. I now regret acting like this with him. Maybe if I had acted normally, he would have told me about the situation in Kupwara. But back then, thoughts that seemed crucial to me—about how I was still paying my dues to the uncle who lent me money to attend Neibu's graduation day, how he had refused to buy me a colour television but lent ₹3,000 to our sister when she asked, how he was again going back while I had to take care of our large family—wouldn't leave my heart and mind. At that point, all I wanted was to be a rich man, with no debts—because the only debt I had taken upon myself was for my brother who, I thought, had failed to see what I had done for him. Thankfully, it was going to be over soon, I thought, since by the end of May I would be able to pay it all off. After that, I would only spend what I earned on myself. That was the plan.

The last trip Neibu made to Nagaland was nothing short of momentous. Neibu, in all his uniformed glory, went visiting all his loved ones—from one friend to another, and one family member to another—in his olive-green uniform, commanding a salute from each of his loved ones. He made sure his last visit to Nagaland as a gazetted officer of the Indian armed forces was imprinted permanently in the minds of all the people he respected and adored back home, including those from church and school.

Our sibling Khrisaneisa still remembers the day Neibu wore his full uniform to go to his old school. He had apparently concluded his talk before a class by saying, 'It's not Neibu teaching you the Pythagoras theorem. It's Lt Neikezhakuo Kengurüse teaching you. Friends, also remember to give me a salute.'

He left a part of himself with everyone he loved on this visit, as if he knew something we didn't. When Neibu visited his friend Rokomhalie in his uniform and spent the night at his place, they had a deep conversation about his hopes and dreams. When Roko asked him, 'How come you are on leave?' Neibu answered, 'We are posted in a disturbed place, so we were all asked to go on leave if we wanted to.' How I wish he had given me, too, a hint of this fact during his last visit.

I later discovered that during this conversation Neibu had also spoken about what he wanted to do with the piece of land he had bought in Rüziephema during his teaching days. He wanted to grow teak there. I don't know what gave Neibu the will to see such dreams, the brightest of dreams, in the toughest of hours. While we were barely making ends meet at home, and Neibu was posted in a conflict zone, protecting our nation, he was still thinking of his life back home, figuring out where we would grow teak and sell it, keeping the security of our collective future in mind.

Nimbu Saab

Neibu's trip to Nagaland came to an end on a Monday morning and just like that, he was on his way back to the Valley while I had left for work, not knowing if I would see him again to say a final goodbye. Neibu had carried back the bull-horn brooches handcrafted by me for all his brothers in arms. That, I now feel, is the only memory I can look back at fondly from his last visit. In my defence, I didn't know better. I didn't know he was going to be in the war zone soon, and I didn't know I would not get to see him again. Because of my immaturity, Neibu had left with an empty feeling, it seems, because later, from Kupwara, he wrote me a letter, in which he said: 'I thought you were going to be with me on Monday!'

At the end of April 1999, Neibu visited Srinagar to take his part B exams with his course mates, including his regimental officers, Thapar, Tomar and Rauthela. It was like a mini reunion for the IMA cadets. While they took the exam, the young men also caught up with each other and made the most of their two or three days of freedom before returning to their battalion headquarters—Kigam, a small village based in Kupwara.

To begin with, Neibu was in the Ghatak platoon, which is supposed to be made up of the physically fittest, most mentally robust of them all. This carefully curated group of soldiers comprises one officer, one Junior Commissioned Officer (JCO) and about sixteen to eighteen men—that is, a total of around twenty invincible soldiers of the Indian Army—who are trusted with missions that

'Your Own Ease, Comfort and Safety Come Last'

are the need of the hour. Typically, a Company has 100 soldiers, and five to six of the best-performing soldiers from every Company make it to the Ghatak platoon. For this reason, they are given separate weapons (LMGs) and the best radio sets, among other reinforcements, to help them carry out special tasks in an operation. Neibu's Company commander suggested that he was probably made to command this platoon because he was physically fit, astute and had a lot of knowledge about the mountainous terrain, a significant advantage for an officer during mountain warfare. Owing to his upbringing in Nagaland, although the mountains there were quite different from the ones in Kargil, he was agile and had the stamina needed to climb the mountains here. Only when Neibu couldn't make himself available for some reason for special tasks was he replaced in the Ghatak platoon. Like when he took his last leave before the unexpected battle of Operation Vijay, Lt Rauthela took over from him during Counter-Insurgency (CI) operations during the time Neibu was on leave.

As I read more about the regiment Neibu served in, I learned that 2 Raj Rif had a long history of winning some of the hardest battles fought by Indian soldiers, including in both world wars. The Rajputana Rifles, raised in 1921 as part of the British Indian Army, is not only the oldest and senior-most rifle regiment of the Indian Army, but its officers have also received innumerable gallantry awards over the years. While this piece of knowledge filled me with pride, I imagine it must have filled the newly commissioned Raj Rif boys with unprecedented bravado and even unimaginable courage.

Nimbu Saab

Even while Neibu was in Kupwara surrounded by terrorism, he never forgot about us. His letters from the war zone made that amply clear to all of us. He would write to me, advising me on how I should think about my career. In a letter dated 8 June 1999, this is what he said to me:

> Remember that there are lots of people who with equivalent qualifications are doing much better than you. It's not because you cannot do it, but it is because you have never tried. I want you to come up in life.

At that time, owing to my misplaced anger, I failed to decipher from these words that Neibu was really just trying to look out for me. My ego probably prevented me from looking at the bigger picture and giving the benefit of the doubt to him.

Whatever the reasons may have been, the fact is that on Neibu's last visit, the only quality time I spent with him was at dinner. I had assumed, perhaps like any brother would, that I had the rest of my life to reconcile with him—a regret I will take to my grave, I suppose.

Coming from Kohima, also a mountain city, a highlander himself, Neibu was one of the most mindful officers in Kupwara. Subedar Sanjay Singh, Subedar Satya Pal Singh and the rest of the men from Neibu's platoon followed his instructions to a tee. They say, Neibu knew the terrain and conditions by heart. He truly led from

'Your Own Ease, Comfort and Safety Come Last'

the front, his men could trust his call on the various routes and climbs in Kupwara. He would simply say, 'This way will be more convenient; *jaldi pahunchenge. Yeh mitti mazboot hai, aur yahaan par dhanss jayega* (We'll reach soon. This sand is firmer here. If we take the other route, our feet will get stuck in the sand).' And his men would follow.

That time in the Valley was such for soldiers like Neibu that if there are ten units deployed in the area, five of them would lose at least one officer. Hence, the feeling among officers deployed in the valley reflected the idea of 'carpe diem'—live your life to the fullest now since anyway the future is unknown. I wonder what that must have been like for Neibu, who had been a strict budgeter all his life. He would write in his diary an account of every little item or action his personal funds were spent on during each day— listing things like a phone call made back home or to Caroline, a movie ticket and 'miscellaneous expenses'. His relationship with money was perhaps sacrosanct because we had not seen much of it growing up. He would often use the '*Pop Goes the Weasel*' nursery rhyme to speak about money.

> A penny for a spool of thread,
> A penny for a needle,
> That's the way money goes.

I sometimes feel deeply saddened by the fact that Neibu's 'save now, spend later' philosophy when it came to money kept him from living his life fully. He spent his youth trying to earn just enough money to take care of our family, and when he was finally in a position to spend on leisure from time to time, he had been posted in a place where not much could be done with that cash. Still, to me, and I'm sure to him too, it was all God's plan—at least until it felt like the plan was truly flawed.

In April 1999, when spring was in full swing in the Valley, the men of 2 Raj Rif had their days filled with strenuous activities. Most mornings started with patrolling to identify terrorism in the neighbouring areas and, some evenings ended frustratingly with terrorists slipping away as they sensed the threat of soldiers approaching them from all sides. The terrorists slipping away on several occasions in Kupwara led to a lot of stress among the soldiers. However, 2 Raj Rif was in good hands, with some of the most determined officers of the Indian Army. Hence, the hope to win all battles, while making sure his men were truly prepared and acclimatized with the terrain overpowered every other feeling or fear in the world for the Commanding Officer (CO), Lt Col M.B. Ravindranath.

Naik Jairam, one of the men whom Neibu served with in D coy, recounts how a tricky seventy-two-hour operation in Kandigaon, in Kupwara, where Neibu's Company was posted, was brought to a halt as control was passed on to BSF jawans because another operation needed 2 Raj Rif's attention more urgently.

Soon after this happened, the men of 2 Raj Rif received deployment orders for Drass, Kargil, as the country needed its army to come together there to protect the citizens of India.

'Your Own Ease, Comfort and Safety Come Last'

A plan that General Pervez Musharraf, the tenth President of Pakistan, famously known as the architect of the Kargil War, had launched in November 1998 was taking shape. Pakistan had successfully declared a war in Kashmir against India, while the protectors of our land—the Indian Army—did not even have the slightest hint of suspicion. Thankfully, a group of shepherds tipped off a few Indian officers posted around Kargil. Without this crucial information the Kargil War would have had a very different outcome—a disastrous one, perhaps, for the whole country. The information passed on by the locals was that they had seen groups of 'mujahideens', allegedly with weapons, dressed in civilian clothes climbing up the mountains along the Line of Control (LoC) in the Indian state of Jammu and Kashmir. From the little knowledge I have, I know that it's unethical and even illegal[1] for soldiers of any army to disguise themselves as civilians to start a war. For this reason, when these 'mujahideens' were caught and killed by Indian soldiers, their bodies were not accepted by the Pakistani Army, perhaps for fear of the shame that would accompany the bodies of soldiers, for the country that was already making news with this infamous move. It was only after the Indian Army sent a couple of patrols from 4 Jat regiment that the need to declare Operation Vijay was felt by the Indian government. Atal Bihari Vajpayee, the then Prime Minister of India, did not formally declare war but announced military action in response to the Pakistani incursion during the Kargil War in 1999. During a conversation with Anil Tipnis, the Indian Air Force chief at the time, Vajpayee reportedly clarified that while India would not allow Pakistan to have its way, 'We will not cross the Line of Control' either.[2]

Nimbu Saab

The first two patrols from 4 Jat, led by Lt Saurabh Kalia and thereafter Lt Amit Bhardwaj, played a crucial role in finding the Pakistani Army, which was stationed on the Indian side of the LoC waiting for the Indian Army to discover them. However, the Indian Army paid a heavy price, in terms of physical and mental torture endured by the prisoners of war, Lt Kalia and his men, because of the intelligence failure in the early days of this conflict.

Operation Vijay was finally made public on 3 May 1999 in Drass, Kargil, located around 16,000 feet above sea level—an altitude where a common man would find it hard to breathe after walking just a few steps. It was here, a stretch of about 250 kilometres that consisted of many peaks, ranging from 16,000 to 20,000 feet, that the Kargil War played out from May to July 1999.

By the time Neibu's D Company got the news of the declaration of the Indo-Pak war, some soldiers had already heard on the radio that 200,000 troops of the Indian Army had been mobilized across the country, and no less than 30,000 soldiers were deployed to regain control of the Srinagar–Leh Highway, Kargil and the Siachen glacier. The Indian Army had been given a tough task: to go for defence after the enemy (the Pakistani Army) had pulled off an extreme, offensive move. In the winter months of Kashmir in 1999, the Pakistani Army broke the unsaid rule of 'not crossing the border' when the border was unpatrolled due to bad weather conditions. When nobody was looking, the Pakistani Army had taken a favourable position on many peaks. It was ready with ammunition, ample ration, administrative bases for transportation and communication, and soldiers; starting from 'no man's land' it had stepped a little into the Indian side of Kashmir. What would have been my peace-loving brother Neibu's reaction to this shocking news? I'm sure he didn't mean war when he told his friends back home about being posted in a 'disturbed area'.

The Indian Army conducts large-scale war exercises annually in different terrains across the country to test operational readiness and adaptability to various combat situations. Such exercises are only meant to make sure that the soldiers of the Indian Army are trained for war-like scenarios if they were to occur. However, the first-ever counter-insurgency training, which all the newly commissioned officers posted in the Valley had been through during the past couple of months, had actually prepared them for the real war-like situation that had emerged out of nowhere. All the cordons, ambushing, patrolling and terrorist encounters in Kupwara seemed like a cakewalk now compared to the news that had hit the Indian Army on the morning of 3 May 1999.

Considering the impossible situation in Kargil, 2 Raj Rif received deployment orders for Drass, and they bid Kupwara goodbye for an indefinite time. The move unfolded in two installments.

First, Thapar and Maj. Acharya accompanied by Havaldar Sharvan Singh and Naik Jairam, amongst other soldiers from all Companies, were sent to Kargil as an advance party. They left Kupwara on 24 May 1999. After staying the night of 25 May at the Srinagar transit camp, the men reached Sonmarg. The rest of the battalion joined them the next day. Meanwhile, the men had set up tents to welcome the rest of the regiment for acclimatization. Without acclimatization, basic survival would have become difficult for the men. For acclimatization, they established a camp in Meenamarg, just ahead of Sonmarg. Ideally, the time to acclimatize for the tough road ahead of them would have been around fifteen days according to the CO, but the soldiers barely got four or five days to get used to the weather conditions before they got orders for their first attack, which had to occur on 12–13 June.

Since this was the only lull period the soldiers of 2 Raj Rif knew they would get in a long time, they made the most of it. During one of their excursions to Sonmarg market, Maj. Vivek Gupta, Maj. Acharya, Tomar, Thapar and Neibu had a good conversation as they devoured what was possibly their last unhurried meal together. Only Tomar had carried his wallet so he paid ₹450 for it and all the others promised to pay him back later—a debt, he says, forever etched in his memory.

After Meenamarg, the next base for the 2 Raj Rif men to continue their training was Matayen, a village located around 21 kilometres away from Drass. The Raj Rif camp was strategically located, in the sense that you could get a sense of all the action without being scrutinized by the enemy. Adjacent to the Raj Rif camp was also a helipad, frequently buzzing with activity due to VIPs arriving at the 8 Division headquarters there (where all the war planning happened), sparking widespread speculation of what was going to happen next on the battlefield. Furthermore, the convoy of the General Commanding Officer (GOC), Maj. Gen. Mohinder Puri, was parked in close proximity. He was regularly spotted traversing the NH1A in a Gypsy, distinguished by its protruding radio antennas and camouflage nets. Interestingly, 8 Mountain Division was formed over fifty years ago by the Indian

'Your Own Ease, Comfort and Safety Come Last'

Army to combat insurgencies in the Northeast. Fast forward to 1999, a boy called Neibu, who was born amidst the insurgencies in Nagaland in 1974, was fighting on the Indian side.

The days leading up to the attack were routine and boring, yet exhausting. Every day, they would climb up a mountain with their rucksacks only to return for some firing practice followed by weapon cleaning.

Until 2 Raj Rif was given their first objective, their days looked long. The men would wait in anticipation of what was to come their way. They'd hear loud firing and observe casualties and martyrs from neighbouring regiments being carried down periodically. Sometimes, when the firing was too loud, or during the daytime in general, they would avoid discovery by the enemy by folding up their tents and taking cover behind bushes close by.

In a nutshell, for the men of 2 Rajputana Rifles, the days leading up to the first attack were filled with nervous anticipation. They also noticed from their location how the brigade headquarters (56 Mountain Brigade commanded by Brig. Aman Aul) based in Drass was a target of the enemy's gun. The sounds of exploding shells and the firing of machine guns were both audible and visible to the Raj Rif men. Every time they would see a fallen Indian soldier, the voice in their heads that said 'They are killing our men!' would become louder. Every passing day of such misery made them more and more restless for their day of vengeance.

To fill up their days of waiting, the young officers of 2 Raj Rif had even set up a music band, which consisted of Neibu on the guitar, Rauthela on the drums, and Tomar and Thapar on the congo and vocals. Some of their favourite songs to play were *'Neele Neele*

Nimbu Saab

Ambar Par', 'Chalte Chalte Mere Yeh Geet Yaad Rakhna', 'Lemon Tree' and *'Lucky Lips'*. Some days they sang, other days they spent their free time in bed revising dialogues from movies like *Top Gun* to gear up for the action they had been looking forward to. On one such day, Neibu captured a moment on Thapar's Canon camera when he, Maj. Vivek Gupta, Acharya and Thapar were sitting in Thapar's tent on his bed. A last photograph that returned from the war with Rauthela and his heavy heart, just without most of the other joyous men in it, including the photographer who clicked it.

Knowing that Neibu had such happy and joyful moments, despite the high-pressure and dangerous situation he was in, in the last few days of his life often gives me some solace about missing out on time with him.

By the time they finally got orders from the top for the attack, the men of 2 Raj Rif were more than ready to defend their country and take revenge for their fallen brothers.

It was 1 June 1999, when 2 Raj Rif received moving orders for Moghulpura. The preparation for the attack of 12 and 13 June started in full swing, and the seniors pumped up their juniors to get them to go above and beyond in the service of the nation they called home.

I'm sure this feeling of unity, of being one with India, would have been new for Neibu. I can imagine the contradictory emotions that he would have felt, unlike most other young soldiers of the regiment who were born in a completely different environment. I'm glad that Neibu got to experience the sense of patriotism that most of us can only talk about.

'Your Own Ease, Comfort and Safety Come Last'

It was during this time, I suppose, that Neibu would have written back to his friend, Lt Vishal, who, during that time, had a peace posting with 6 Dogras (an infantry regiment) in Dera Baba Nanak in Gurdaspur, Punjab. Vishal used to make fun of Neibu for picking one of the apparently safer arms, ASC (Army Service Corps). Alluding to the fact that it was one of the service corps, Vishal, who always wanted to join the infantry would tell him, 'You must hide behind an *aate ki bori* (sack of flour).' Now that the tables had turned, Neibu joked with his buddy saying, 'I'm going to Drass, Kargil and things are getting heated up here. You always wanted combat. Where are you now?'

10

Bare Feet Feel the Earth's Heartbeat

TOLOLING WAS ONE OF the most crucial points in Kargil. It was under the control of the 6 Northern Light Infantry (NLI) of the Pakistani Army. It gave the enemy a vantage point to see everything on the other side. Thus, even vehicles carrying rations for the Indian soldiers would sometimes get hit by enemy fire. In addition to that, the Pakistani Army had infiltrated deep into the Indian side and had captured an area of over 127 kilometres—4 to 10 kilometres deep into the Mushkoh Valley, Drass, Kaksar, Kargil, Batalik and Turtuk sectors. Such deep intrusions by Pakistan meant that they could occupy posts located on the Indian side of Kashmir at any moment, especially the posts overlooking NH1A—a highway that connects J&K directly with the rest of India. This could be a disaster in the making for India if it were not taken care of by the Indian Army.

Tololing became the turning point of the Kargil War and, unfortunately, the reason behind the loss of some of the bravest soldiers of the Indian Army. It was the site of an extremely important operation for the Pakistani army, named Op. Badr (also called Koh-e-Paima, meaning 'the one who climbs'). Besides, it was the pet project of General Pervez Musharraf, because of its strategic location.

The events of the Kargil War, from the point of view of the Indian Army, were concentrated in the Tololing area and was defined by its significant landmarks. The Tololing ridgeline itself featured prominent points like Tololing Feature, Point 4590 and Tololing Top, with a level stretch known as Flat between them, followed by the Barbad Bunker. Progressing northward from there was a landmark known as Hump, beyond which a string of smaller hills stretched, sequentially named from Hump 1 to Hump 10. Further north stood the area's most commanding elevation point, Point 5140.

Now, to launch an assault on Tololing from Drass, the Indian Army had two principal options: one route was via the south-western ridge, leading directly to Point 4590, and another ran along the south-eastern ridge approaching Tololing Top. Before the task of launching this assault was passed on to 2 Raj Rif, the task of clearing out the Tololing area was given to 18 Grenadiers, supported by 16 Grenadiers and 2 Naga. The repeated attacks, which were unsuccessful due to abysmal intelligence on the situation, led to the Indian Army losing some of their valiant officers, like Maj. Rajesh Adhikari and Lt Col Ramakrishnan Vishwanathan, among others. In this attack, India had very expensive losses, including a MiG-21 as well as an Mi-17 helicopter flown by Wing Commander Rajiv Pundir with six other soldiers, who were all shot down by a missile just above Tololing. Understandably, during this time, the morale of the troops was down and difficult to manage.

3 June 1999: This was the day when the role of 2 Raj Rif in the Kargil War was decided. The battalion was given the task of capturing Tololing because of the previous failed operations there, and they were assured of the support of the artillery regiments, which were well-established in that area by then.

On 12 June 1999, A Coy and D Coy were sent for the first attack led by 2 Raj Rif. At this time, Neibu was the Ghatak platoon commander, and Rauthela was the officiating coy commander of B Coy. Unlike other regiments that had been in the area for a few days, the men of 2 Raj Rif had entered the war with fresh pairs of eyes and unparalleled motivation as well as determination.

Conquering Tololing was no easy feat, and it took multiple attempts for the Indian Army to finally hoist the tiranga (the Indian tricoloured flag) at its peak. The multi-directional attack included C Coy, under the command of Maj. Vivek Gupta, who was tasked to lead the assault from the south-eastern approach and capture Tololing Top and Flat. B Coy was to act as the reserve for this phase of operations. The H-hour for this attack was 8.30 p.m. Simultaneously, from the south-western approach, D Coy under the command of Maj. Mohit Saxena was tasked to capture Point 4590 and Rock, including Barbad Bunker, with A Coy as reserve. With a difference of thirty minutes, the H-hour for their attack was 8 p.m.

The idea for the multi-directional approach was to deceive the enemy about where the main thrust was coming from and divide enemy reaction.

At exactly 7 p.m., the pre-H-hour firing by the mighty Indian artillery began. They unleashed hell on the enemy. Taking advantage of the disruption created by the Indian artillery firing and the chaos on the enemy's side, Maj. Gupta's C Coy marched ahead from the south-eastern side, and D Coy, with Maj. Mohit Saxena in the lead, attacked from the south-western side. As per plan, the enemy was deceived and had re-sighted most of their weapons towards the south-western approach, thereby leaving some room for C Company to exploit the situation. While in the face of heavy enemy resistance, Maj. Saxena's D Company had to stall the attack from the south-western approach.

During this battle, Neibu and his Ghataks established stops for the capture of Tololing. While Saxena was attacking Point 4590, he had an interesting exchange with Neibu and Maj. Bajaj, one that Saxena looks back at fondly till date. Both Neibu and Saxena were placed in opposite directions and were supported by the same fire base that was controlled by Maj. Bajaj. Now, there were occasions when the base used to fire either close to the location where the Ghataks had taken up positions or close to Saxena's assaulting Coy. So, every time that the gunfire would come close to the Ghataks, Neibu would tell Maj. Bajaj, '*Sir, aap mere upar kyun fire kar rahe ho?*' (Sir, why are you firing at me?) On the other hand, whenever the fire fell short of its objective, Saxena would say, '*Sir, ab aap mere upar fire kar rahe ho.*' (Sir, now you are firing at me.) Bajaj would have at least one of them complaining to him during the time, putting him in a fix! This is now a memory that both him and Saxena hold close to their hearts.

Meanwhile, Maj. Gupta's Company went ahead and captured Flat and eventually, Tololing Top, by 12.30 a.m.

This unfavourable situation for Charlie Company, which was unable to go for the attack according to the strategy, led to a change of plan, which came from 'One Five' (code name for 2 Raj Rif's commanding officer). The officer gave to the Alpha Company, which was in reserve for the attack from the south-western approach, led by Maj. Padmapani Acharya, a chance to unleash their men on the enemies in the south-western approach, a.k.a. the Barbad Bunker. Thapar led this attack with Maj. Acharya's orders while the rest of the Company gave him cover from 200 metres behind. It was a steep climb, and the boys were not as acclimatized as they would have liked to be, but one loud war cry of *'Raja Ramchandra ki jai'* was enough for them to charge towards the enemy bunker, especially considering they were so close to their goal. It was not only the war cry, I believe, that led their spirits to soar again, it was also the anger at losing their brothers that gave them that extra boost and stamina for the attack. This particular attack was nothing like what they had seen before that day—there were Bofors, mines, shelling, constant firing from both sides. Pakistan lost eleven soldiers in this attack, and 2 Raj Rif lost a dozen soldiers during this first attack. Though the attack was successful, it came at a very heavy price. On 13 June 1999, 2 Raj Rif also lost its first officer, Maj. Vivek Gupta, with Tomar barely escaping a medium artillery shell, which left every soldier in the Company shaken. Now, although Tololing Top was captured and so was Flat, the enemy presence on the Feature was still intact because of the unsecured gap between Point 4590 and Tololing Top.

Neibu reported in his journal after this attack:

> After much planning and recces done at various places (including recce by Coy commanders), the attack was launched on the night of 12–13 June on Tololing, overviewing the entire Drass area.
>
> The attack was launched with 18 Grenadiers, providing a firm base to the attacking Coys. C and D were the two

assaulting Coys, with B and C as reserves, respectively. D Coy, under the command of Maj. Mohit Saxena, attacked from the south-western direction and captured trig height 4,590. C Coy, under command of Maj. Vivek Gupta (Lt Praveen Tomar took over after the fall of Maj. Gupta), attacked from the south-eastern direction and captured Tololing, which was the main objective of the unit. The two reserve Coys later married up and helped in capturing the entire Tololing Feature.

The Gathak pl., which was under my command, went from behind the enemy. We positioned ourselves to cut off the reinforcements of the enemy.

Though my party has not faced any real attack like the other Companies, it was a hard task. Fortunately, the enemy did not open fire from the Helmet top, as they were pinned down for most of the time by arti shelling and other supporting weapons. Or else, I may not have been writing these lines now.

The attack was a successful one, and has opened up a way for the Indian Army to proceed further.

At the wreath laying function for Maj. Vivek Gupta and the rest of the soldiers who had sacrificed their lives for the country, the men of 2 Raj Rif seemed restless to take revenge for their soldier-brothers. As the survivors in the attack, including Neibu, Thapar, Tomar and Acharya, rested and recouped at their base post in Matayen, it seemed like they had lived fifty years during that one evening of battle.

The price that 2 Raj Rif paid for Tololing was expensive, but attaining victory on that peak was needle-moving for the rest of the battle for the Indian Army.

Several days after the conflict, acclaimed journalists made their way into the tents of the soldiers who had just returned from the Tololing battlefront. This was the war that gave birth to real-time reporting in India as journalists took to keeping the nation informed about the ongoings at our borders. I don't know how they did it but despite the exhaustion they must have faced, soldiers from the Kargil War found it in themselves to give a detailed and vivid account of the battle's events to the young, enthusiastic war journalists, as they shared with them a serving of rum.

I would not have been able to believe then if someone told me that Neibu was witnessing his soldier-brothers breathing their last, crying for their families as they lay there in blood, in pain perhaps—and sometimes with parts of their bodies missing. When you're at war, there's no time to mourn, there's only little time to get your men together, more united than ever before, pick up martyred bodies, collect your enemy's weapons and ammunition, clear out the enemy's scattered bodies after a victory, and most importantly, gather courage and resilience for the next attack. The Indian Army soldiers would often find letters of Pakistani Army soldiers while clearing out the points captured, and with every letter, more secrets about the planning of this disruptive war would be revealed. At the same time, they would empathize with the soldiers writing letters to their loved ones, as they themselves had done the same during troubling times, desperate as they often were to make their families feel at ease, since the news of the war had spread like wildfire by then and it told a different story than what most soldiers were reporting to their families in their initial days. However, as the attacks got intense, soldiers started speaking the truth, as did Neibu, since there was no guarantee that they would live to tell these tales to their loved ones.

After Tololing, Neibu also called home and spoke to Apuo honestly. He said, 'We have won the battle and snatched many arms and ammunition of the enemy, but many of my loved ones have fallen in front of my eyes and I am not feeling good about it.' Despite the grim news, our ever-optimistic Apuo assumed that this meant that after the impossible Tololing win, they would now be given some relief and someone else would take charge in future attacks. Nobody in my family knew then that the infantry and artillery worked hand in hand during this war, and many regiments were acting together to change the course of the future in favour of the Indian Army—which seemed like a distant dream on day one, owing to the unfair advantage that the Pakistani Army had. Therefore, our father felt a little relaxed instead of worried after the call, completely unaware as he was of the gruelling challenges and relentless battles that lay ahead for Neibu and his comrades. He really had no clue that his son, as the second-in-command after Maj. Mohit Saxena, was now gearing up for the next attack.

The battalion received fresh orders within two weeks of the first attack, on 24 June. 2 Raj Rif was to attack the Three Pimples,

Knoll and Lone Hill areas—all prized possessions for Pakistan and India. While Three Pimples was a contiguous feature of three sharp and steep peaks west of Tololing Nala, Knoll and Lone Hill were located around the Three Pimples ridgeline.

Without the collaborative strategy of the sub-units and units within a battalion, and a carefully designed dance between artillery firing and the infantry's on-foot soldiers, any attack during Operation Vijay was incomplete and impracticable.

Therefore, again, strategically, the tasks were divided between Companies. D Coy, with Maj. Mohit Saxena in command and Neibu as the second-in-command, was given the responsibility of capturing Lone Hill, and A Coy, led by Acharya and Thapar (as the second-in-command), was given the responsibility of Knoll.

The attack was to happen on 28 June, and the recce for it began three to four days in advance. As a part of this objective, Neibu had to do the recce of the objective—Lone Hill—from the east side. During the recce, Neibu found a wire that was running from Lone Hill to a feature called Point 4700, most likely a communication line between two enemy Companies. This discovery of communication at the right time enabled D Coy to cut the conspicuous line on the day of the attack, successfully targeting enemy capabilities.

On the night of 27 June, the batallion started their climb and reached the fire base by 6 the following morning. Just then, heavy enemy shelling killed one of their men, Satpal. The firing continued sporadically and so did the briefing on how to mindfully proceed with the attack going on.

Amidst the chaos, Maj. Saxena, Acharya, Thapar and Neibu shared a light moment and had a conversation about home and family. On the evening of 28 June, the officers sat down at the firm base known as Kajal, and my brother told his crew about the large family he had waiting for him at home. I learnt about all of this much later, of course, through Saxena. Some of the officers were hearing about Neibu's large family back home for the first time and felt a surge of compassion and pride in him for his sense of duty towards his family and his determination to change the trajectory of his life. The origin of his sense of duty, which they had only been observing in the tasks assigned to him so far, became clear to them now. Here was a man hardwired to uphold his responsibilities to the best of his ability and then some.

They also caught up on some sleep. Finally, around 8 p.m., the men of the D Coy started their climb towards their objective. Before these young, angry men could digest the fact that the enemy had already killed one of their dear soldiers, Satpal, they lost many more soldiers in the unexpected shelling that ensued. One soldier, Havaldar Jagdeep, was hit by a splinter, which pierced through his helmet. Many of their weapons were also damaged, the fire base was disrupted, and so, before the determined, audacious men of 2 Raj Rif could begin their task, they saw it all break down in front of their own eyes. From where they were standing, they could see the Three Pimples complex clearly, but from where the enemy was standing, they (the enemy) could point a gun exactly at anyone they chose. At this point, even a rock rolling down from the top could be a powerful weapon for the enemy.

In such unthinkable times, there are only two things that work for a soldier—faith and astounding courage. Especially in this second attack for 2 Raj Rif, the only choice the men had was to pick up the pieces of their broken hearts and morale and make the enemy pay for the loss of their brothers, their buddies, their squad, their leaders. What helped them were the words of the 2 Raj Rif commanding officer, Lt Col Ravindranath, who in their sainik sammelan earlier had shaken them out of their sense of loss and pain, and said sternly to them, 'You are the undefeatable Rajputs and Jats, and you have to take the Three Pimples complex from the Pakistanis. This day is what we join the army for, and today the country needs us to SHOW UP and we have to prove our mettle.' These words were enough to wake up the sleeping tiger in each soldier's heart.

D Coy found three peaks, known as 'false crests', on their way to Lone Hill. They passed through false crest 1, where Neibu went with his platoon and found it empty. Next, they went to false crest 2, which Saxena went to with his Company and which also turned out to be empty. They then decided to make false crest 2 their fire base. Thereafter, the remaining hill and their objective to be captured was Lone Hill, for which Neibu started—with his Platoon 10—at about 2.30 a.m. on 29 June. Saxena, meanwhile, stayed at their fire base and continued engaging the enemy.

At the same time, since Knoll and Lone Hill were on the same ridgeline, Knoll, in the forefront, was being attacked by A Coy. Lone Hill was at the back and was being infiltrated by D Coy. Fire base 1 and 2 (which were ahead of Kajal) were trying to engage the objective, Lone Hill, which was upwards of 1,000 metres away.

After A Coy's attack from the front, D Coy was supposed to come from the back and attack the enemy. This way, the enemy would be caught off-guard and would have nowhere to go. However, as D Coy started stepping into the 'Nala', a bomb fell near the artillery OP (an artillery personnel responsible for directing the fire in support of attacking troops) and the soldier got injured, as a result of which, D Coy had to leave two of their men behind to take care of the soldier. To make matters worse, their Regimental Medical Officer (RMO), Capt. Basu, also got injured in this attack and had to be left behind as well.

After sacrificing many men at the fire base itself, the soldiers of D Coy were struggling to find a ray of hope in this operation. With a fair amount of grit, they moved forward, keeping in mind the goal of capturing Lone Hill. Naik Jairam was with Neibu in Platoon 10, who was tasked to lead the attack on Lone Hill.

From the west of the ridgeline, A Coy had kept the enemy busy, which gave the D Coy men a window to climb up from the east. On the way, D Coy navigated minefields to make it to their target without any more casualties. They couldn't afford any more casualties and deaths if they wanted to attain victory.

However, despite the fact that they had a window of opportunity open in front of them, Neibu, Naik Jairam and the rest of Platoon 10 realized that there was an obstacle they had not accounted for during this attack—it was an almost vertical (70–80 degrees) rock-like incline that had to be climbed to reach the point where the enemy was. There's a reason, I guess, why this peak is called 'Lone'

Hill—it is indeed an isolated hill, far from the main mountain ranges of the vicinity.

The only silver lining in this situation was that the enemy was sitting at an acute angle, so if Neibu's platoon could manage to climb up, defying all the odds, the enemy would not be able to fire at them directly. The angle of the slope they would be climbing would ensure that any fire from the enemy would miss them. For once, after a long time, the unwelcoming non-navigable peaks of Kargil had favoured the Indian soldiers a tiny bit. Therefore, the boys of Platoon 10 of D Coy were absolutely not going to give up this golden opportunity. In this war, these were a rarity, so you had to take advantage of what nature had to offer you in this terrain.

While on this side, the Lone Hill men were trying their best to climb up—their only chance to capture the Three Pimples complex—on the other side, Thapar and Acharya of A Coy had been martyred while capturing Knoll.

It's hard to imagine now but apparently, during the attack, Thapar walked right into his ultimate sacrifice after realizing his mentor Acharya was no more; he yelled at the enemy, '*Bhag jao, kaffiron, tum kya ladoge* (Run away, you infidels, what will you fight)?' The enemy on the other side responded, '*Ruk jao, saalon, tumhara to bhagne ka time nikal chuka, ab janaaza hi niklega* (Wait, you bastards, the time for you to run away has passed. Now only your funeral will leave from here).' Unfortunately, this curse from the enemy turned into a reality within seconds for Neibu's brave and cheerful friend and band member, Thapar. No amount of platoon member Hav. Dhan Singh's pressing his head down in front of the enemy helped Thapar, as if he had decided in his head to die fighting. For him, it was a dream to win a gallantry medal from this war, and win he did albeit posthumously.

On the other side of this agonizing series of events in A Coy, while Thapar's head rested in the arms of Tilak Singh, D Coy, unaware of anything that had happened on Knoll, were midway up Lone Hill. At one point, the climbing got dangerous, as the surface was slippery and hard, making it impossible for the soldiers to find any grip and move forward. Not knowing what to do and refusing to descend from this point, when they had already made a lot of progress, I don't know what came over Neibu. He commanded his men to take off their shoes, hang them on their shoulders and help each other climb up. Being a true leader, Neibu was the first to take off his shoes despite the bone-chilling air and the freezing surface of the stone before him.

Once the twelve or thirteen soldiers of Platoon 10 of D Coy made it up, they were able to take advantage of the position of the enemy, as their line of sight was clear now. Neibu signalled to all his men to 'stay quiet', with a finger on his lips. They had not come this far to give up. Neibu wanted to make sure that the enemy did not discover them, so they could move ahead with the operation with the element of surprise intact. As they approached the top of the hill, they started noticing the silhouetted figures of the enemy and along with flashes. Jairam was in the eighth or ninth position, carrying an LMG, with Riflemen Tatra, Hanuman and Sharvan Singh ahead of him. By the time Jairam reached the second stretch of the incline, there was a distance of 10–15 metres between him and Neibu. The men of D Coy were confident that if things went well, they would reach the top, and since they could then overpower the enemy with their position, they would catch hold of them before the enemy could realize where they came from.

Jairam even began saying , '*Daboch lenge*' (We will take them down) in his head as they were progressing, to remind himself of their goal. Just as his thought was about to turn into reality, one of the men slipped. Neibu and the others opened fire instantly, breaking the silence and giving away their secret operation. The fight was an impossible one for the Indian soldiers. However, taking charge of the situation, Neibu and Sharvan Singh fired a few rounds at the enemy and the men from the other side started falling one by one, screaming, '*Allah Hu Akbar* (God is the greatest).'

At this point, a point of no return, Naik Om Prakash of Maj. Saxena's platoon came up with the idea of screaming out war cries in different Indian languages to perplex the enemy and give them the impression that a strong and fairly large army of Indian soldiers was coming for them. Therefore, all the soldiers from the fire base started shouting in different directions, with different war cries such as '*Raja Ramchandra ki Jai*' and '*Jo bole so nihal*'. Maj. Saxena's platoon hoped to psych the enemy and give them a signal that a large army was coming for them from either of the sides. However, it was too late. The acute angle that worked against the enemy before had now turned in their favour as the unstoppable soldiers of 2 Raj Rif had almost climbed up the steep slope. So now, the enemy could see everything clearly, and they came down with full force on Neibu's platoon, who had managed to pull off an extraordinary act of bravery right before this unfortunate moment. First, they rolled rocks off from the sangar onto the men climbing up, and second, they opened fire with their AK47s and grenades. Their fluke shots from up above caught my brother, Neibu. The bullet hit his abdomen, which then started bleeding profusely. However, he marched forward to lead his men, by example, to finish what he had started. His men asked him to rest but his only response was, 'I'm okay, I will tighten my belt and the bleeding will stop.' As he approached the top of Lone Hill, he shot straight at the enemy bunker and neutralized the first one. Because of this the enemy from the next bunker grew alarmed and started firing at them. Neibu and his men fired back in retaliation and killed two more attackers.

Unfortunately, just as the brave men of 2 Raj Rif were about to attain victory, my wounded, exhausted brother was shot in the right eye which led to him falling off the top of Lone Hill. Neibu, in what must be the darkest hour for the Kengurüse family, took a fall of 200 feet. He had done it in the name of his nation, the nation he had grown to feel much closer to than ever before. Along with him, Sharvan Singh sacrificed his life, while Hav. Lal Singh's arm was blown to bits. Naik Bhavani and Hav. Jaipal also got badly injured.

Platoon 10 of D Coy had lost a leader but not his resilient spirit. The fear that was triggered in one enemy bunker because of Neibu, trickled down to all the other Pakistani soldiers on Lone Hill. The firing continued from both sides, with the count of injuries and deaths going up on both sides as well. In the end, the men could see the enemy running away while screeching '*Allah hu akbar, humein mat maaro* (Don't kill us)' in pain. In the middle of this chaos, Naik Jairam also got shot. The bullet in his thigh, he said, felt like a metal coin heated to the highest temperature possible. While the enemy ran for their lives, Neibu's daring men roared, '*Nimbu saab, ye aapki jeet hai, ye aapka hill hai.* (Nimbu sir, this win is yours. This hill is yours.)'

I wish my beloved brother had survived for just fifteen minutes more to witness this hard-earned triumph. I've thought about this a lot over the past few years, and I've often wondered: Did Neibu hear his men declare this win as his before he took his last breath? I hope and pray he did.

Parallelly, after the loss of Thapar, an angered soldier from his platoon, Shivnayak, gathered his men and went ahead with the unforgettable attack of 2 Raj Rif. Knoll and Three Pimples were fully cleared by 9 a.m. on 29 June. But the battle was long over for my brother Neibu and his dearest friends in uniform, Thapar, Acharya and many of the men they commanded on the battleground—all the men who were once photographed together by Neibu in Thapar's tent. In the photograph, their boredom is apparent on their faces. When this boredom would relent to the extent that it would turn them into martyrs in the Kargil War, I suppose, none of them could have ever guessed.

Now, in heaven, they must be singing '*Sandese Aate Hain*' (Messages keep coming in) together when they see their men hoisting the Indian flag, dancing, claiming the land that was always theirs.

After a few years, I got to know from a course mate of Neibu's that during his second term at IMA (January–July1998), my dear brother had volunteered to perform a song on stage for one of the social evenings.

Nimbu Saab

Neibu, who had always been a good singer, put together a dramatic solo performance that left the audience in awe. The song was inspired by a song from the Bollywood movie, *Salaami*, where the son sings it for his father. Neibu cycled to the stage, holding the mic and singing '*Bachelor Boy*', which went as follows:

> When I was young, my father said,
> 'Son, I've got something to say.'
> And what he told me I'll never forget until my dying day.
> He said, 'Son, you'll be a bachelor boy,
> And that's the way to stay.
> Son, you'll be a bachelor boy until your dying day.'
> When I was sixteen I fell in love
> With a girl as sweet as can be,
> But I remembered just in time what my daddy said to me.
> He said, 'Son, you'll be a bachelor boy,
> And that's the way to stay.
> Son, you'll be a bachelor boy until your dying day.'
> As time goes by, I probably will
> Meet a girl and fall in love …
> Then, I'll get married, have a wife and a child,
> And they'll be my turtle doves.
> But until then, I'll be a bachelor boy,
> And that's the way I'll stay.
> Happy to be a bachelor boy until my dying day …
> Yeah, I'll be a bachelor boy,
> And that's the way I'll stay.
> Happy to be a bachelor boy until my dying day.[1]

Neibu, despite all the love he had in his heart for Caroline, indeed died a bachelor boy. But he didn't forget to reveal his love for her to Apuo in a letter. He had already asked for her hand and told us to take care of her if he didn't come back.

Almost all of Neibu's friends (except Tomar) had written down their last letters for their loved ones, leaving the letters with their buddies in case they didn't return. Neibu wrote us all a letter too, but it was only in his journal—which he left for us as a kind of playbook to help us go on with our lives without him.

2 June 1999

My beloved brothers and sisters,

I don't know if I'll be able to meet you again. In case I don't come back, read this letter together and please fulfil these requests:

1. Love and respect our parents, and always obey them.
2. Love one another, for if you love one another, no one can ill-treat you.
3. Continue to keep your faith in the Lord and strive for your spiritual growths.
4. Love your friends and avoid bad company.
5. Give your best in your studies.
6. The younger ones—respect your elders.
7. The elder ones—love your younger ones.
8. Never break your bond with each other.
9. Live an honest and pious life.

10. Never fight among yourselves.
11. Do not hate your enemies nor persecute anyone.
12. If I have done any wrong to any of you, please do forgive me.
13. As for me, I don't have grudges against any of you.
14. Please be kind to my girlfriend; she has been my rock and support till the end.

It has been twenty-five years, and we're still trying, to the best of our capabilities, to follow his commandments. As a family, we do wonder sometimes if we've made Neibu proud yet—the answer, unfortunately, is always 'No'. I reckon that's a burden every war martyr's family lives with.

Epilogue

It wouldn't be wrong to say that almost every member of the Kengurüse family carries the burden of being a part of the family of a hero. The difference, however, is that while, to the world, Neibu may be a hero because of the ultimate sacrifice he made in the line of duty, to his own family, he was a hero whom each of them needed and depended on. Yet, the one overwhelming feeling that the Kengurüse family unanimously shares is that somehow they do not live up to the high standards set by Neibu. Even though all the Kengurüse siblings are achievers in their own right, they can't help feeling that they have somehow fallen short of what they could have achieved if Neibu were around. Almost every frame adorning the wall of the living room of the family home in Kohima speaks loudly of the glory of Capt. Neikezhakuo Kengurüse, a Maha Vir Chakra recipient, and the pride that they feel in their hearts. It seems that both in life and death, he gave them extraordinary dreams to hold on to. And if that's not what truly defines a hero, nothing else can.

His father, however, carries along with this pride a deep-rooted pain about the life his second-born led. The pain is not just about losing this son at a young age. It emanates from a regret, which he carries till date and possibly will for the rest of his life: the regret that his relentlessly dutiful son couldn't spend even one day simply savouring the joys of life, without being riddled by pressures. He

Epilogue

grew up in poverty, and his adulthood was spent pushing hard against his circumstances, paving the way for a better future for him and his family, and just when he seemed to have made it, he left the world without even enjoying the fruits of his labour. What did his near-perfect son think about in his last moments? Did he even get a chance to think? Nieselie will perhaps never find the answers to these questions, and his eyes will perhaps never stop carrying the vacuum that his son's death left behind. So, while he continues to attend many state events and be felicitated for his son's achievements, it is never not accompanied by the hurt that he silently carries in his heart.

The news of Neibu's martyrdom arrived on 29 June 1999, and during the days following its arrival, it shook Nagaland the way no other event had—except, ironically, the death of the Naga nationalist leader Zapu Phizo, also sometimes known as the father of the Naga nation, in 1990. Most of Neibu's friends and extended family heard about his fate in Kargil during an Angami programme over the radio. One can imagine them sitting and listening to the news, scrunching up their faces while perhaps increasing the volume to make sure they were hearing the name right—the name they were hoping to not hear. Neibu's immediate family, however, which was scattered across Nagaland, found out from other sources.

As luck would have it, a little before that most unfortunate day, Neibu's grandfather had been really sick. Both Nieselie and Atoulie were in Nerhema to take care of him. So, on the afternoon of 29 June, while Nieselie stayed with his father in the village, Atoulie went to Kohima to get some medicines. They had spoken to Neibu

Epilogue

just a few days ago when he had informed them about succeeding in the last task that they were given, that of capturing Tololing. The news had relieved them as they assumed that since they had had a win; his battalion must be resting while another one takes over, they thought. That's how much they understood about wars. They really thought they had no reason to fear anything.

Thus, barely two weeks later, when Atoulie was waiting at the Kohima bus stand to take the bus to Nerhema, clutching his grandfather's medicines in one hand, and he saw his cousin's husband, Kameswar Singh, walking towards him, gesturing to him urgently to come listen to him, his fears centred on his grandfather. Neibu's well-being wasn't even a distant concern. But when he came closer, his brother-in-law blurted out the words '*Neibu murishe koishe*' (Neibu is dead!). For a minute, Atoulie couldn't make sense of anything. All the noise around him seemed to have died down, and he stood there, numb and a little angry with his brother-in-law for delivering such terrible news to him in the way he had. However, in the next minute, as he gathered his bearings, he convinced himself that something must be wrong. This news couldn't be true. They all knew that Neibu had already finished the last operation safe and sound. This had to be false news. His brother-in-law told him that he had caught the news on TV and now Atoulie was being summoned to the Bethel Hospital in Kohima.

How could Atoulie accept this news? He hadn't even had one normal interaction with Neibu after his passing out parade. In fact, on Neibu's last day at home, Atoulie had been so angry with him that despite knowing that Neibu would be returning to his battalion

Epilogue

soon, Atoulie had left for work that morning without even bidding his brother farewell. He had done it deliberately, knowing very well that Neibu was expecting to say bye to him in person. None of Neibu's letters from the battlefield had been answered by him, either. He hadn't cared for Neibu's efforts, perhaps assuming that they had all the time in the world to settle things between them.

On his way from the bus stop to the hospital, Atoulie's belief that it was false news turned into prayers. 'This cannot be true' and 'this is false' soon became 'let this be false, please let this be false' as he reached the hospital. Soon, as he looked at the faces staring back at him, he knew that all his prayers had gone unanswered. The reality stared stubbornly, unforgivingly in his face.

In the meantime, the news had reached his other family members as well. His father had found out through some army officials who had reached Nerhema looking for Neibu's family home, while his stepmother found out through a phone call. The call informing her had been made on their neighbour's phone in Jalukie, since they did not have a phone of their own. The only thing they knew was that Neibu had given the supreme sacrifice but none of them had any idea how events would unfold from that point onwards. So, when his stepmother went home after the call, despite having to keep wiping the tears that were streaming down her face, she simply started preparing a spot for Neibu's mortal remains to be kept when they would be brought back, which she had assumed was going to happen on the same day.

The most striking thing for the family, which was used to living in quiet anonymity owing to their humble background, was that suddenly, everyone around them not only claimed to know

Epilogue

them, but claimed to be closer to them than they had ever been. In fact, as soon as it was decided that Neibu would be brought to Kohima, there was more than one 'uncle' who declared that the coffin should be brought to their place. It was as if overnight, the Kengurüses were the most famous family of Nagaland.

The family found out that Neibu's mortal remains would arrive in a week. With a family as large as theirs, organizing even the smallest of moves was a tall task. Atoulie and his father decided that it was thus best for the family to gather in Kohima. The number of people wanting to pay their respects to Neibu was increasing every day, and the house in Jalukie was too small to accommodate that kind of crowd. Besides, there was no suitable place in Jalukie for the same.

Thus, the next few days were spent trying to gather the family in Kohima at a relative's house. But Atoulie realized that it was also on him to let two people know: his birth mother and Caroline. He requested his aunt Medolhouü, who worked in the same department as his mother, to make sure that she was informed. For the latter, he decided he would go himself.

On the fateful morning when she received the news, Nupulhuu was in the kitchen, putting a pot on for their tea. The morning hadn't really taken off to a great start since she'd had some disturbing dreams on the night before. First, she said she dreamt of someone rolling downhill in front of her eyes, which made her wake up with a start. When she did manage to sleep again, for some reason she dreamt of Neibu standing at her doorstep and calling out to her. But before she could respond, her eyes opened, and her son vanished. She now stood looking at the boiling tea,

Epilogue

her heart feeling as uneasy as the liquid gurgling in front of her, when she heard a knock at the door. Could it be Neibu like in her dream? she thought. When she saw a peon with a paper in his hand instead, she was confused. When he asked her if she was Capt. Neikezhakuo Kengurüse's mother, her heart sank. She realized how the dream she had woken up from last night had been the worst nightmare of her life. The peon informed her of Neibu's martyrdom and also told her that he would be flown down to Dimapur soon.

Nupulhuu stood at the doorway, looking out the door long after the peon had left. Her world had come crashing down once again and this time, she couldn't even see her son's face. She couldn't believe that she had lost her second born too. How scared she had been for him. Her mind kept going back to the conversation she'd had with Neibu when he informed her of his decision to join the army. She had asked him: What is the need for taking up such a dangerous profession when you seem to be doing so well as a teacher? But he had convinced her that with this choice he would be able to do much more for his siblings and that he was really happy to have been selected. This made Nupulhuu take a step back and let him enjoy his moment of glory, despite the ache she had felt in her heart. That ache was now back and seemed to have increased manifold. Only now, it was incurable.

Caroline, on the other hand, found Atoulie standing at her doorstep early one morning. The news of Neibu's sacrifice had spread throughout the city, like wildfire, so she had already heard. But Atoulie knew Neibu would have wanted her to be treated a certain way, and Atoulie was not going to leave undone anything that he knew his brother would have wanted. Caroline invited him in and asked him to sit down, quietly moving away to get him a cup of tea. As Atoulie held the cup in his hand and Caroline looked at him, words seemed to fail both of them. After spending a few minutes in much-needed silence, Atoulie finally spoke up. He said, 'You and Neibu may not have married, Caroline, but we are aware that you both loved each other. For us, you are family. I just want to let you know that should you want to come when his mortal remains arrive and see him, or anything that you may want to do to get your closure and pay respect, we would be happy to hold that space for you. We are with you.' As Caroline broke down, with pain and gratitude, Atoulie placed a hand on her shoulder to comfort her and quietly left her to grieve.

All of Neibu's family had been brought together in Kohima by this time. They had not even had the chance to see the remains of Neibu, but Atoulie had to already start filling Neibu's shoes. The more people found out, the more work around Neibu's funeral arrangements piled up. This meant that Atoulie would leave in the morning and only get back to his family at night, after a full day of meeting and making arrangements. Many of his siblings were still quite young and so, leaving them all to themselves in Kohima, at a time when everyone was dealing with the loss in their own way, would worry Atoulie. It was, after all, his responsibility now to make sure that they were all taken care of. So, every morning, he would sit with them before leaving, and starting from the second youngest, task each of the siblings with the responsibility of taking care of a younger one. In this way, he would make sure all of them were looking out for each other.

Epilogue

At this tough time, his biggest blessing turned out to be Aseü. Even though he didn't reach out to her on his own during this time, Aseü took it upon herself to make sure that his family members, especially his young siblings, were taken care of. She would come over to their relative's house in the morning and only leave at night after settling each of them down. When Atoulie would call to thank her after returning home, tired from a full day of running around, she would ask him to not worry about anything and just make sure that he was taking care of himself too.

When Lt Rauthela arrived with Neibu's mortal remains, it was a week since Neibu's family had been living with the knowledge of his sacrifice. To say that the state of Nagaland transformed into one big family on that day would be an understatement. Even the chief minister of Nagaland had come to receive Neibu. The airport was swarming with hundreds of people who had gathered to welcome the son of their land and accompany him on his last journey. Streets were lined with more people wanting to catch a glimpse of their hero. Posters with the words 'Welcome, son of Nagaland' were displayed at the airport. No one witnessing the events on that day could ever believe that there ever was an anti-India feeling amongst the people of the state. The love, respect and honour with which they were awaiting the arrival of Neibu, who had laid down his life while protecting India, was awe-inspiring.

That is exactly what Rauthela felt when he saw the sheer numbers in which people had turned up. He too had been instructed by his friends and peers to 'be careful' while going to the state, but what he saw there changed his feelings about it within seconds. There was no denying the fact that each person standing there was only

Epilogue

and only proud of their hero and his act. Rauthela realized he had nothing to be scared of. Only a lot of adoration and respect to soak in. It was much beyond his expectations and something he knew he was going to carry with him for the rest of his life.

Atoulie, on the other hand, did not have much opportunity to soak anything in, since he sensed an impending problem when he saw his birth mother's face in the crowd at the airport. Since Nupulhuu had been informed of Neibu's arrival, her grieving heart had made it impossible for her to stay away from her son. Her husband and children understood it well. Besides, they too were mourning the loss of their stepson and stepbrother, whom they had grown to be so fond of and who seemed to have been rudely snatched away from them by the hands of destiny. So, Nupulhuu had arrived at the airport with her husband and two of her children, Kezeneilhou and Keneianheii, in tow. When Atoulie laid his eyes on her, Nupulhuu was almost on her toes, her neck stretched upwards as far as it could while her anxious eyes darted from here to there as she looked for signs of Neibu's arrival and walked slowly through the crowd.

Warmed by the sight of her, he rushed to be by his mother's side before there was any chance of an interaction with his father and gave her a warm hug. Realizing that her husband and children had also accompanied her to pay their respects filled him with love, and he looked at all of them with deep gratitude.

A part of Atoulie wished that he could bring her forward as Neibu's mother but understanding the sensitivity of the situation and knowing that it was well understood by her family too, he looked on while her husband gently nudged her to take a step back

Epilogue

and let Atoulie and his father handle the officials accompanying Neibu's coffin.

From the Dimapur airport, Neibu's mortal remains were taken to the Bethel Hospital in Kohima before being transported to the Assam Rifles centre. At the hospital, when the family, along with hordes of other people, gathered around the coffin to pay their respects, it was suggested that since it had been so long since his death, and as they had reason to believe that Neibu's remains may not be in a state suitable for the family to see, that the coffin not be opened. Everyone knew and remembered Neibu in their own way, and so, in order to protect each one's memory of him, Atoulie and his father accepted the suggestion. There were so many people, including their cousins, Neibu's friends and colleagues, who had come over from all around to pay their respects to him, Atoulie said. So many that he could not make note of who all he could see, let alone meet anyone personally. Besides, his eyes were constantly on the coffin that now housed his beloved older brother, finally sleeping peacefully. He could not get himself to look away even for a whole minute.

Eventually they moved to the Assam Rifles centre in Kohima to conduct a guard of honour for Neibu. In the vast, open area, in the middle of the centre, Neibu's wrapped and decorated coffin was placed. Some officials and people from the family stepped up on the podium to talk about him and his sacrifice. Rauthela too decided to speak. Now that he had seen and been a part of the grand and heartfelt way in which the people of this beautiful state were sending their son off, all the apprehensions that had been sown in his mind by the instructions and words of caution back home seemed to melt away. He realized how, at the end of the day, whichever state you may belong to, when you stand at the border of your country to protect it from its enemy, you are only an Indian. He stood on the podium and spoke about the state of Nagaland, the Angami tribe and how brave their son was and how proud every Indian was of him today. When he looked around he realized, with the emotions visible on the face of every young boy and girl in the crowd, which had been tirelessly following them since they stepped out of the airport, carrying posters and shouting slogans, that a big shift had happened. A big group of young men broke into a war cry (which was basically a high-pitched howl) that was reserved only for the bravest warriors of their land, followed by 'Neikezhakuo, *miche tuo heluo*' in Angami, which can be translated as: 'Don't be scared on your way, Neikezhakuo, you are a real man!' Rauthela could sense that though Neibu may have been an ordinary boy from a humble family in life, in death he had achieved extraordinary respect and purpose and caused a definite shift in the youth of his state.

His aunt Medolhouü said she wished she could meet Neibu and say, 'Thank you so much, Neibu, for you have done a big thing for us. As mentioned in the Bible, if a seed does not fall down into the earth and die, how will a tree come out of it? You, my son, were the seed that lay into the earth and led to all the prestige and honour that our family gets now.' It wouldn't be wrong to say that Neibu made these words ring true for every person of the state who, till date, carries the pride of belonging to the place that Capt. Neikezhakuo Kengurüse came from.

Epilogue

Before he passed away at the age of twenty-five, Neibu had bought a small plot of land in Phezha, the place where his birth parents had got married. It was decided that he would be buried in the same place. After the guard of honour, when his coffin reached the place along with his family, the narrow uphill road leading up to it was lined by people for kilometres—not one dry eye in sight. Everyone, including Rauthela, couldn't help but break down as Neibu was laid to rest with the kind of respect and honour that no other person in his hometown was going to be bestowed with for years to come.

When the ceremony was over, Atoulie looked at the grave, gently taking out from his pocket his sweet Neibu's last letter, which had reached Atoulie just before his mortal remains had reached them. Neibu had finished the letter with 'Why don't you write to me? I am your loving brother'. Atoulie knew that he would never get over the regret of losing his brother, his confidant, before he had let him know how much he loved him. He read his last words for the umpteenth time, after which he folded the letter again and kept it back in his pocket, tears streaming down his otherwise calm face.

Almost every citizen who witnessed Neibu's last journey remembers it vividly—as if it happened yesterday. There isn't a single man, woman or child born in Nagaland in the years after Neibu's sacrifice who wouldn't know of Capt. Kengurüse. It was Neibu who showed the youth of his state not just a very legitimate option of a career path, he also gave them a glimpse of what it meant to

Epilogue

don the uniform and do your job honestly and to the best of your ability, even if it entailed taking a bullet for your country.

He may not have survived to see the goal he had for his family—restoring their status and honour—come to fruition, but like they say: Some goals are so worthy, it's glorious even to fail.

A Note from Neingutoulie Kengurüse

HIS UNEXPECTED BUT COMMENDABLE patriotism aside, all Neibu ever wanted for our family was for it to be respected and find a place of dignity in society. Due to our unfortunate beginnings, Neibu wanted our family to thrive, and so he made sure we did. All those years ago, when the relationship between the Indian Army and Naga people was still fragile, he did what no one expected him to do as a Naga. It was not that he had no choice; he had other offers and opportunities. He could have been a regular office worker with a nine-to-five job and still be able to support his family well. Yet, he took a risk, a risk that cost him his life ... Neibu had nothing to assure himself of the rightness of his decision; no money, no support from the elders initially, no backing from a higher authority. All he had was his unwavering faith in God. He boldly charged ahead as though God had promised him greatness in person.

While his heroic actions are well-known and even recorded in history, people who never met him may never know him as a person, as a man with flaws and as a genuine human being. Yet, Neibu truly touched the lives of his friends and family and those who knew him personally in his short lifespan. He wasn't perfect; as his brother, I know he wasn't ... but he was close.

While Neibu could not see his family during his last moments nor see his family's situation improve, his death and his sacrifice

A Note from Neingutoulie Kengurüse

for his country started a chain of events within the family as well as in his people. Had he not done all that he did, maybe today our family would not be able to walk with pride and honour, like we do today; had he not gone out there with nothing on his back but his faith in God, my faith would not have grown. I would have still been a high school teacher and would have stayed where I was twenty-five years ago. I would not be confident enough to venture out and meet people, I would still be in Nerhema living a quiet and unknown life. I found out a bit too late that he had refused to give me a TV because he didn't want me to get comfortable with my life in Nerhema; he wanted me to strive to do more. He was looking out for me all the way.

Thanks to Neibu, I was able to get a petrol pump allotted to us with some effort after my interaction with the defence minister, as mentioned in the prologue, and it did enable me to take better care of our family. Our parents living in comfort and respect, our siblings having his support and sharing his blessings even beyond his death … none of these things would have happened had he not decided he would join the Indian Army.

It was not just our family that he left an impact on but the whole nation, especially us Nagas. Strangers cried for him when his body landed in Nagaland, people praised him, and he restored the public's faith in the Indian Army. He was a man who just wanted the best for his family, and in his attempt to secure his family's future, he gave so much to his country.

If I ever get a chance to meet him again, even for a moment, I will say, 'Neibu, my dearest brother, today, because of you I am a businessman who is doing well for himself and his family. Maybe

A Note from Neingutoulie Kengurüse

it's because of you that God granted me the desire to share my story and never forget a single detail. I do that now with my three children (with Aseü, whom I married), all of whom have grown up listening to your stories—not just the heroic deeds, but even the parts that made you human. I share the stories that made me look up to you, the parts where you taught me, the bits where you were my friend, when we laughed and cried, and even down to the very moments when we fought. I guess anger runs in our blood, we lacked wisdom in our youth, and maybe because of that the last memories I have of you were bittersweet—too many things were left unsaid. If I had a chance, I would tell you how much you meant to me, how because of you and your sacrifice you restored the faith I had once lost in God. Today, if I can be brave, it is because of you. I see your life as the embodiment of "unfailing faith". To the world you are remembered as a hero, a martyr, but I will and always remember you as a big brother, a filial son, a loving friend and most important of all, a man of God. Hence, I also need you to forgive me. I'm sorry, Neibu, for not writing to you. Please forgive your ignorant little brother.'

Through this book, if I were to give any message or piece of advice to the younger generation, or anyone as a matter of fact, I would say that Neibu was just an ordinary man, not a genius or a saint, but he chose to be the exception. He chose to be hardworking, he made the best of whatever he had and even when he didn't have enough, he still chose to share. He gave his anxiety and fears to God and simply walked in faith. He disciplined himself not because he was talented or gifted, but simply because he chose to.

Unintentionally, he left memories and traces of himself in all the people he met, worked and lived with. That is why they still remember him despite his demise twenty-five long years ago. In his entire life, all he did was choose, for he knew well that the true power he had was over his choice. His circumstances were not an excuse for him. He chose not the smooth-sailing road but the difficult and bumpy one, for he knew what lay beyond. He was sure of it ... and he was right!

Acknowledgements

I, Neha, write this on behalf of Diksha and me.

When I was thinking of one word to describe our collective experience of writing this book, what instantly popped up in my head was the word 'layered'.

Writing about our heroes became a part of our lives from the year 2016, when we attended the Kargil Vijay Diwas celebrations for the first time. When we met the families of other heroes, we truly began to feel the power of their stories and also realized how we were all connected. As they say, 'same-same but different'.

Therefore, now that Diksha and I have finally come together to do one of these stories as co-authors, we must begin by thanking the first person who gave us both the push we never knew we needed … Priya Ramani. I can't help but wonder if we would ever have started this journey without her.

While both of us have written books about heroes in the past, *Nimbu Saab* wasn't just that for us. Here was a hero we both had heard of during the course of our past projects (*Letters from Kargil*

Acknowledgements

for Diksha and *Vijyant at Kargil* for me), and we had felt a sense of intrigue about and attachment towards Neibu. During my research for *Vijyant at Kargil*, his name would invariably come up, and I often thought, how did this quiet and genial Naga boy, whom everyone spoke so fondly of, find enough courage in himself and the will to scale the heights he did, in the manner that he did? What made him give such little regard to his personal safety, I wondered. On the other hand, whenever Diksha was asked about whose letters from her book were her favourite, aside from our Daddy's of course, she would answer without batting an eyelid: 'Capt. Kengurüse's'.

So, when the opportunity to find out more about his personal story came up, it was clear that there was no subject better-suited for the both of us to come together as authors for the first time. We both felt equally invested to know more about this legend. This is the first layer of our experience.

The other layer came in the form of the state to which he belonged. Neither of us knew what to expect. What the culture of the Nagas was like, how they were as people ... we knew very little. The only time that we had spent time in the Northeast before this was when we were in Assam in 1991–93 while our father was posted there. Interestingly, Diksha was given her name there, after some trial and error. We were just kids then and didn't have many experiences outside of those with our family and our parents' friends. So, Nagaland was completely new territory for us. But we knew we had to explore it well to be able to understand the story of Neibu.

Personally, it also gave me a chance to reconnect with the lovely, gracious people of 2 Raj Rif, who as always were very welcoming and

Acknowledgements

helpful, despite us troubling them again and again. We must thank the commanding officer, Col. Kardia, for making the visit to the regiment comfortable and fruitful. We are also very grateful to Sub. Babu Lal Chaudhary, Sub. Satya Pal Singh, Sub. Ishwar Singh and Sub. Sanjay Singh for sharing their memories with Neibu and giving us valuable insights into his role as their leader. Brig. Mohit Saxena, Col Praveen Tomar and Lt Col Sanjay Singh Rauthela, we owe you so much for taking the time and effort to explain various aspects of not just Neibu's journey but also of the war to us. They have all helped us a lot. A special thanks to Naik Jairam for telling us what no one else could have, since he personally scaled the mountains of Kargil with Neibu. He was able to remember and share the names and hometowns of the people with whom he went to war.

We all know how big a part the academy plays in shaping the officer. Hence, there was no way we would have been able to access that side of Neibu's life had it not been for his course mates Lt Col Vishal Bais Singh, Col Sundeep Khatri and Col Sachin Duseja.

Coming back to the layers of our experience, the most fulfilling one was getting to know Neibu's beautiful home and his people. Our trip to Nagaland was an experience that we will always carry with us very fondly for the rest of our lives. We must start by thanking Atoulie. He was our narrator, our Naga dictionary and our humorous right-hand man who began to fondly call us 'sis', and we can confidently say he treats us in a manner that befits the word. We gained a brother in him, and rumour has it, he makes a great one. Jokes aside, we cannot thank him and his lovely wife Aseü enough, for being our family and home while we were there. A big thank you to his father Neiselie, his stepmother Dinuo and all the other siblings, whom we got a chance to interact with—Khrisaneisa, Neiketuü, Arhenuo, Mhozienuo, Neichunüo and Viketoulie. Thanks to them for agreeing to relive their pain and share their tears with us. We will never forget the surreal moment when we were all done and ready to leave, and Neibu's father told us he wanted to pray for us. The whole family joined him in doing so. We still get teary-eyed when we think of it. Our hearts were bursting with the love we received and will always hold it safe. Like I said, we had no idea what to expect, and we couldn't have gone back feeling more fulfilled than what we eventually did.

We are also grateful to Neibu's birth mother Nupulhuu and her family, especially her youngest daughter Keneinuo, who drove us around happily, chatting with us, answering all our questions, despite her fear of driving uphill from Dimapur to Kohima. We can't imagine how difficult Nupulhuu's journey in life has been. Yet, when we asked her if she would be willing to go over it all again for us, she did it all willingly and with a smile, even though she had to stop sometimes to wipe the tears that kept forming in the corners of her eyes. Her voice still echoes in our heads. '*Ab main kya bolega?*' (What can I say now?) she said, but anyway went on to narrate to us her side of the story. We remember how we sat in her home in Dimapur, feeling restless, helpless, for the cards she had been dealt with.

Neibu and Atoulie's childhood friends—Rokomhalie, Thepfuvilie, Ravannuo, Akhrenuo and Sanuo—whom we now know were the reason why Neibu had a fun-filled and beautiful childhood and adult life, sat with us enthusiastically and patiently shared the fond memories they'd had with Neibu. They would

Acknowledgements

answer all the follow-up questions we had. His aunt Medolhouü and all his cousins—Kiyasetuo, Thepfusanuo, Solelhoulie and Hebuo sat with us in their respective homes, often sharing a meal, talking about Neibu. In that moment, it was clear they hadn't forgotten a thing about him and they were gracious enough to bring those instances to life for us to be able to write this story.

Among many others, Neibu's colleague Kepelhoutuo and his beautiful wife hosted us in their lovely home and shared their memories with us with such warmth.

We are truly moved by how almost every single person that we met holds Neibu so dearly, and we definitely feel the regret of never having had the chance to meet him. Through this journey of interviewing Neibu's loved ones, we came to be in awe of how Neibu's character was perceived in exactly the same way by everyone we met. We were quite impressed and honestly, a little sad that we would never get the opportunity to know Neibu in real life.

But the fact that he fought the war alongside our brave father, Maj. C.B. Dwivedi, a Sena Medal awardee, connects us in a way that nothing else can. Even though Daddy isn't here with us, not a day passes by when we don't think of him and wish that he is looking at us, and feeling some of the pride that we hope he is. It is all for and because of you, Daddy, our first and favourite hero.

On the army side, we owe our thanks to Lt Col Vivek Tripathi from ADGPI for his generous support during the entire process. This wouldn't have been possible without him.

We are also very grateful to our agents and our editorial team, or as Diksha likes to say funnily, 'our partners in passion', who have been with us from start to finish. Our literary agent Kanishka and his teammate Narayani helped us immensely with our book proposal, and then the wonderful team at HarperCollins, especially our publisher Swati and our editor Anju took over. We tested their patience sometimes, but they were always so lovely and hands on with their inputs and, more importantly, constructive feedback, which made us think harder, and hopefully, write better. This was not an easy book to edit by any means.

Our mother, Bhawna Dwivedi, must be thanked for her support, patience, love, encouragement and the many other ways in which she comes through to help us, for we would be nowhere without her. To say that she's always our best inspiration to keep shooting for the stars is an understatement. All it takes for us to keep going in a difficult situation in life is her voice in our heads saying, 'If the thirty-two-year-old ridiculously pampered housewife Bhawna, despite losing her loving husband suddenly, could, you can.'

This acknowledgement cannot be complete without thanking the third child of the Dwivedi household, Lt Col Rohin Chhibber for helping us figure out the parts we couldn't, communicating on behalf of us at times and for absorbing all of the anxiety and impatience through the process. You didn't have to, but you always do. Thank you for all the love and support you shower upon us, despite the annoyance we bring to your life.

To our wonderful circle of family and friends who are often our loudest cheerleaders and on the rare occasion are even our most lenient critics: Thank you for pushing us, and pulling us up, from time to time, and for always knowing which of the two we need at any given time. We must especially mention Brig. Ajay Dalal, for being readily available to not just answer our queries but also

Acknowledgements

going to great lengths to authenticate the responses even without us asking. We really value you and your friendship.

Thank you to every single person we're close to. From the bottom of our hearts, we thank you for putting up with us while we were writing this book, knowing well that writing a book on a war that your father fought is tougher than it seems. We're just glad we found the strength and courage to tell these stories from time to time, and for that, we would like to thank our respective support systems, which oftentimes is also just Diksha for me and vice versa. Hence, also thanks to each other, I guess.

I should also thank my dear friend Faraz for many reasons, the least of which is him helping me with the contract of this book. Diksha's childhood friends, Akshat Kumar and Param Tripathi, have known for years when and how to show up for her, and they always do it. Thank you, Akshat and Param for that and for helping us with the MOU even before Diksha and I knew we would be writing this book together.

Lastly, thank YOU, our wonderful family of readers, for picking this book. Your love has and continues to encourage us to keep moving on, one step at a time, one story at a time. No stones have been left unturned to do justice to this project, considering the complexities of the subject; we hope you appreciate the finished product.

Again, thank you for picking this book. Authors can merely open doors, but it's the reader who steps inside and breathes life into what are essentially just scribbles.

Notes

1. Sense of Belonging

1. Sayak Basu, 'History of separatism in the conflicted northeastern state of Nagaland', *Deccan Herald*, 26 February 2023, https://www.deccanherald.com/elections/history-of-separatism-in-the-conflicted-northeastern-state-of-nagaland-1195265.html.
2. 'The 16 Point Agreement between the Government of India and the Naga People's Convention', UN Peacemaker, https://peacemaker.un.org/sites/peacemaker.un.org/files/IN_600726_The%20sixteen%20point%20Agreement_0.pdf.
3. V.R. Krishna Iyer, 'Saga of the Nagas', *Economic and Political Weekly*, 29, no. 12 (1994), https://www.jstor.org/stable/4400958.
4. *Ibid.*
5. Visier Meyasetsu Sanyu, 'Growing up in the jungles of insurgency hit Nagaland: Visier Meyasetsu Sanyü's Naga odyssey', *The Caravan*, 17 November 2018, https://caravanmagazine.in/conflict/excerpt-naga-odyssey-long-way-home.
6. *Ibid.*
7. *Ibid.*
8. *Ibid.*

4. Three Minus One

1. Binalakshmi Nepram, 'Drugs, guns and four stories from the Northeast', *Outlook*, 25 September 2020, https://www.outlookindia.com/national/india-news-drugs-guns-and-four-stories-from-the-northeast-news-360878.
2. A gospel camp is a get-together organized by the Christian community. It is organized for the youth, and, in Jalukie, it usually takes place once or twice a year.

7. Naga GC Creates History

1. Command tasks are tasks assigned. They usually involve a problem to be solved alone or with a team. The idea is to test your skills as a team player and leader.

8. The Rise of Lt Neikezhakuo Kengurüse

1. A punch is a cocktail of different spirits like rum, whiskey, gin, etc. It is usually poured into a sizeable glass or mug.

9. 'Your Own Ease, Comfort and Safety Come Last'

1. As per international humanitarian law under the Geneva Conventions.
2. 'Vajpayee refused permission to IAF to cross LoC during Kargil conflict: Tipnis', *The Times of India*, 28 June 2019, https://timesofindia.indiatimes.com/india/vajpayee-refused-permission-to-iaf-to-cross-loc-during-kargil-conflict-tipnis/articleshow/69994773.cms.

10. Bare Feet Feel the Earth's Heartbeat

1. '*Bachelor Boy*' is a song by Cliff Richard and the Shadows, written by Richard and Bruce Welch (from the Shadows).

About the Authors

Neha Dwivedi is an alumnus of DPS RK Puram and Lady Hardinge Medical College, Delhi. She is the founder of Happy Birth-day, a brand for a team of birth workers dedicated to empowering new parents through the pregnancy, birth and post-partum phases. A daughter of a Kargil War martyr, she sought and found solace and strength in writing.

Neha is the author of *The Lone Wolf: The Untold Story of the Rescue of Sheikh Hasina* (2021) and *Vijyant at Kargil: The Biography of a War Hero* (2020). *Nimbu Saab* is her third book. She aims to continue working towards bringing forth the stories of heroes like her father for she believes that they all deserve to be heard. She would love to hear from her readers and can be reached at nehadw@gmail.com. Her Instagram and Twitter handle is @nehadw.

Diksha Dwivedi is an author, entrepreneur and seasoned storyteller, with a deep-rooted passion for empowering narratives. She has an economics degree (2009-12) from Shri Ram College of Commerce and a master's in journalism (2012-13) from Cardiff University. YOSO Media and AkkarBakkar.com, two media ventures founded by her, have reached and resonated with millions across the globe. Diksha is the author of *Letters from Kargil* (2017). Through her diverse roles, she continues to inspire a culture of storytelling as a powerful means of connection and change.

HarperCollins *Publishers* India

At HarperCollins India, we believe in telling the best stories and finding the widest readership for our books in every format possible. We started publishing in 1992; a great deal has changed since then, but what has remained constant is the passion with which our authors write their books, the love with which readers receive them, and the sheer joy and excitement that we as publishers feel in being a part of the publishing process.

Over the years, we've had the pleasure of publishing some of the finest writing from the subcontinent and around the world, including several award-winning titles and some of the biggest bestsellers in India's publishing history. But nothing has meant more to us than the fact that millions of people have read the books we published, and that somewhere, a book of ours might have made a difference.

As we look to the future, we go back to that one word—a word which has been a driving force for us all these years.

Read.